Agricultural
Science

Agricultural Science

E. C. Hewitt B.Sc., Dip.Ed.
Chadacre Agricultural Institute

D. J. Brazier B.Sc.
Warwickshire College of Agriculture

Third edition

COLLINS
8 Grafton Street, London W1

Collins Professional and Technical Books
William Collins Sons & Co. Ltd
8 Grafton Street, London W1X 3LA

First published in Great Britain by
Crosby Lockwood & Son Ltd, 1968
Second edition by Crosby Lockwood Staples, 1974
Reprinted 1976 (twice), 1979 (original ISBN 0–258–96958 X)
Reprinted 1980 by Granada Publishing Limited – Technical Books
Division (ISBN 0–246–11355–3)
Third edition published by
Collins Professional and Technical Books, 1986

Distributed in the United States of America
by Sheridan House, Inc.

British Library Cataloguing in Publication Data
Hewitt, E. C. and Brazier, D. J.
Agricultural science.—3rd ed.
1. Science   2. Agriculture
I. Title   II. Brazier, D. J.
502'.463   Q161.2

ISBN 0–00–383279–1

Typesetting by Columns of Reading
Printed and bound in Great Britain by
Mackays of Chatham, Kent

Contents

Preface   xi

Section I   Basic principles
1   Applied physics   2
    (A)   Mechanics   2
    (B)   Electricity   25
    (C)   Heat   46
2   Applied chemistry   70
    (A)   Elements, atoms and molecules   70
    (B)   Formation of ions   72
    (C)   Solutions   74
    (D)   Base exchange   77
    (E)   Carbon dioxide   79
    (F)   Oxygen and oxidation   80
    (G)   Acids, alkalis and pH   82
    (H)   Metal corrosion   83
    (I)   Water and water supplies   83
3   Cytology and inheritance   86
    (A)   Cell structure   86
    (B)   Plant cells   87
    (C)   Animal cells   88
    (D)   Cell reproduction   90
    (E)   Genetics and the work of Mendel   92
    (F)   Sex linkage   95
    (G)   Animal breeding   97

Section II   The plant
    Introduction   101
    Classification of plants and animals   101
4   Plant morphology and anatomy   104

	(A)	Monocots and dicots	104
	(B)	Root, stem and leaf	105
5		Plant physiology	115
	(A)	Photosynthesis	115
	(B)	Water absorption and movement in the plant	118
	(C)	Mineral absorption	121
6		Plant propagation	125
	(A)	Vegetative propagation	125
	(B)	Seed propagation	126
	(C)	Germination	132
7		Plant competition and crop yields	135
	(A)	The maximisation of crop yields	135
	(B)	Requirements for healthy growth	136
	(C)	Weed control	136

Section III   The animal

		Introduction	139
8		Anatomy	140
	(A)	Basic structure	140
	(B)	Joints and bones	142
	(C)	Skin	143
9		The digestive system	146
	(A)	The mouth and the transport of food	148
	(B)	The stomach: simple and compound	150
	(C)	The small intestine and absorption	151
	(D)	The large intestine	152
	(E)	Rumen fermentation	152
	(F)	Chemical digestion: enzyme action	155
10		Blood and the circulatory system	158
	(A)	The constituents of the blood and their functions	158
	(B)	Blood vessels and the heart	161
	(C)	The circulation	164
	(D)	The mechanism of blood clotting	165
11		Animal reproduction	167
	(A)	Male reproductive organs	167
	(B)	Female reproductive organs	170
	(C)	Pregnancy and birth	173
	(D)	Genetic manipulation and embryo transfer	176
12		Growth and maturity of farm animals	178

Section IV   The environment

Introduction   181
13   Meteorology   182
   (A)   Climatic factors and recording   184
   (B)   The value of weather records   200
   (C)   Interpretation of forecasts   202
14   Soil   205
   (A)   Rock formation and constituent minerals 205
   (B)   Soil formation   210
   (C)   Texture, structure and classification   213
   (D)   Effect of soil constituents on soil fertility   215
   (E)   Soil improvement   217
   (F)   Soil water, drainage and irrigation   218
   (G)   Natural water supplies   225

Section V   Beneficial and harmful organisms

Introduction   227
15   Microbiology and principles of hygiene   228
   (A)   Fungi   229
   (B)   Bacteria and viruses   236
   (C)   Spoilage and storage of plant products on farms   241
   (D)   Hygiene   245
   (E)   Useful micro–organisms in industry and agriculture   248
16   Some important invertebrates   252
   (A)   Classification of invertebrates   253
   (B)   Protozoa: single-celled animals   254
   (C)   Platyhelminthes: flatworms   256
   (D)   Nematodes: roundworms   260
   (E)   Annelida: earthworms   264
   (F)   Arthropoda: jointed-limbed animals with hard skin   266
   (G)   Mollusca: slugs and snails   282

Section VI   Chemicals in crop production

Introduction   285
17   Fertilisers   286
   (A)   Principles of manuring   286
   (B)   Nitrogen: nitrogenous fertilisers   291
   (C)   Phosphorus   292
   (D)   Potassium   295
   (E)   Fertiliser usage   296

(F)   Organic manure   298
(G)   Residual value of fertiliser   301
18   Chemical crop protection   302
(A)   Herbicides   302
(B)   Insecticides and acaricides   304
(C)   Fungicides   304
(D)   Reliance upon chemicals   305

List of supplementary reading   307

Glossary of terms   309

Index   315

Acknowledgments

The authors wish to acknowledge the help and encouragement given them by their wives and also the valuable advice and constructive suggestions of Mr. K. C. Vear of Seale-Hayne Agricultural College.

Preface

Early man acquired scientific knowledge from the observation of every day occurrence by experimentation. When books began to be printed and communications were improved, young men were able to obtain basic information from the older, experienced investigators who had preceded them.

Science can only progress in this way, or else intelligent men would spend their lives re-discovering facts established by Archimedes and others years before.

The scientific field today is vast and almost frightening in its scope. No one man can live long enough to become an expert in more than one branch, but that does not prevent him from being interested in other developments.

In agriculture we already depend upon machines, chemicals and a knowledge of plant and animal behaviour to maintain world productivity. We are well aware that our present level of production is insufficient to supply the needs of a rapidly increasing world population; so in future we hope to raise the output of the less productive parts of the world to supply more food where the need is great.

Economic pressure has been the driving force behind much of our agricultural improvement in the United Kingdom. Yields of winter sown wheat crops have been raised from just 2.4 tonnes per hectare in 1950 to nearer 7 tonnes at present. The need to increase financial returns/hectare is forced upon us by the high rate of interest on the capital invested in land, buildings and machinery.

A crop of wheat cannot produce 8 tonnes of grain per hectare if insect pests are feeding on the roots and stem at ground level. The chemical control of soil pests and plant disease has become essential to maintain levels of productivity. Fertilisers are being used to achieve maximum growth potentials in crops, and our meat-producing animals can be fed with the correct amino acids to increase their growth rate.

A basic understanding of scientific principles is necessary to enable the modern farmer to modify his technique in the light of the findings of agricultural research.

The three main scientific disciplines are biology, chemistry and physics. Each is sub-divided now into more specialised groupings. For example, biology includes botany, zoology, microbiology and genetics. Within these groups there are further divisions of study, as within zoology there is entomology, the study of insects, which is a specialised field producing much valuable agricultural information.

Agricultural science is largely applied ecology, or the study of an animal or plant in its environment. The chemical, physical and biological effects of the surroundings on a plant or animal need to be understood.

For this reason information is required from many sources to solve the problems of plant and animal growth.

One often hears agriculturalists say how difficult it is to keep abreast of new techniques. A basic knowledge of science enables one to achieve this more easily. In fact, many modern methods will remain behind a closed door for a farmer with little or no scientific background. In the competitive world of today all young people who have chosen farming for a living must have a knowledge of agricultural science.

The aim of this book is to present the subject in a manner that is easily understandable to those who have not studied science. Those who have 'O' levels in science subjects will understand how the generalities they have learned may be applied to become the specialities of agricultural science.

Section I

Basic principles

Introduction

In order to understand most subjects it is essential to have a good grounding in basic principles. The scientific principles dealt with in this section are those which need to be understood before machinery, animal and plant science can be properly explained.

Complicated chemical and physical occurrences have been described in a straightforward way. For example, electricity or electronics is something to do with electrons. There is in this section an explanation of electrons, how electric current is produced, and used safely with economy. The application of this knowledge, through machines, is easy to see.

All the basic information covered in this section is necessary to the complete understanding of later sections and modern husbandry practices.

1 Applied physics

(A) Mechanics

(a) Measurements

When describing an object scientifically or perhaps recording what has been the result of some trial or experiment, we need to be able to express the dimensions of an object or substance in universally accepted units. Old time agricultural measurements show interesting local variations. For instance, there were nineteen different-sized acres used in Great Britain, ranging from the Leicestershire acre of 1,930 m^2 to the Cheshire acre of 8,562 m^2. Imagine the confusion should a farmer move from one district to another. Units of measurement have now evolved through national systems to the present 'Système International' often mistakenly called the 'metric system'.

Weight

If one is suddenly confronted with the need to know the weight of something, such as a bale of hay, an estimate is made of the weight of that bale by comparing it with other objects whose weight we might have experienced: an assessment the accuracy of which depends very much on skill. Much greater accuracy is achieved by comparing the object of unknown weight with other objects of known weight, using a weighing machine or balance.

Practical difficulties arise when we want to weigh a fat bullock and we have to carry around perhaps four hundred kilos in weights to balance the animal. To record the accurate weights of small quantities in the laboratory, a balance is used and the units are grammes and milligrammes; while on the farm, weighing machines incorporating various leverage devices enable the weight of such things as fat beasts to be balanced by weights of much less actual weight.

Strictly speaking, any object contains a certain amount of material and the weight of the object is the gravitational pull on that amount of material. A distinction is made between the *mass* of an object, which is always constant, and its *weight* which might vary as gravity might vary at different heights and in different places. If a 560 kg cow were taken to the moon, her mass would not change, but her weight would only be 89 kg. This change is only detectable when an object is weighed by a spring balance which measures gravity pull by tension on a spring. For all practical purposes on the farm, the degree of accuracy in weighing things does not demand any consideration of changes in gravity.

Volume

One easy method of measuring quantities, particularly if the material is a standard and familiar one, is to express those quantities as the amount of space occupied. In fact, before weighing machines which were robust and accurate enough for farm use were generally available, volume measurement was universally used for farm produce. Like the acre, the bushel—which was the usual volume measure—varied from district to district; a bushel of wheat varying from 28 to 36 kg. Volume is still a convenient way of measuring quantities around the farm, especially when the material is a liquid. The metric units of volume used are the litre and millilitre. The conventional way of marking glassware to indicate a contained volume is to inscribe a line, which should be level with the lowest point on the liquid surface, to measure the accurate volume.

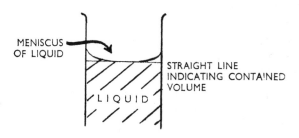

Fig. 1.1 Accurate measurement of liquid volume.

Density

One has only to experience handling a wet bale of hay from a stack to realise that the relationship between apparent size and weight is not a constant one, even with similar material. The weight of material in a certain volume is known as the *density* of the substance concerned. For instance, the density of pure milk can be stated as 1.03 kg per litre, or wheat density could be stated as 1,226 litres per tonne. The interaction of units can be very confusing, and so relative density (or specific gravity) is a much tidier way of recording density. By comparing the weight per unit volume characteristics of any substance with those of water, a simple figure can indicate density of the substance compared with water. This may be found from the expression:

$$\text{Relative Density of a Substance} = \frac{\text{Weight of any Volume of Substance}}{\text{Weight of same Volume of Water}}$$

If one litre of milk weighs 1.03 kg and one litre of water weighs 1.0 kg, the relative density of milk is 1.03, i.e. 1.03 times denser than water. Densities can therefore be expressed on a scale relative to water with that of water being 1: materials with a relative density less than 1 being less dense than water; while those with a relative density greater than 1, being more dense than water, will sink when placed in water. Examples:

Ice	0.92	Aluminium	2.7
Oil	0.9	Lead	11.4
Wood	0.5	Mercury	13.6

The only factor which can influence the density measurement of a pure substance is temperature. Since a mass of material increases in volume as its temperature rises, it becomes less dense since the same weight of material now takes up more space; and so densities are commonly measured at a temperature of 15°C where possible.

The practical use of density is as a means of ascertaining if a substance is pure, since any admixture of an impurity of different density will raise or lower the density reading. For instance, instead of having to carry out a painstaking analysis of milk to check if water has been added, it is merely necessary to check the relative density of the milk; any lowering of the reading indicates added water.

How can relative density be easily measured? Its convenience might be lost if measurement techniques were difficult, but there is a very simple solution to this problem. We have to thank Archimedes for its solution: his famous Principle states that a body immersed in a liquid displaces its own volume of liquid. If the body floats, it displaces a volume of liquid the weight of which equals the weight of the object. The object used to measure relative density of liquids is known as a hydrometer. When it is placed in a liquid, it will float when it displaces its own weight of liquid which is constant. In a less dense liquid it will have to displace more liquid to equal this weight and so will float lower in the liquid. By

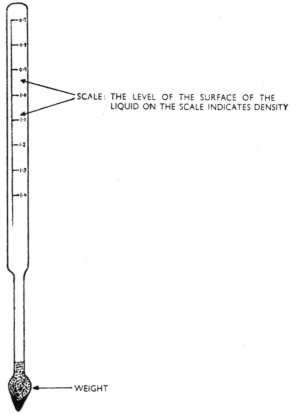

SCALE: THE LEVEL OF THE SURFACE OF THE LIQUID ON THE SCALE INDICATES DENSITY

WEIGHT

Fig. 1.2 Floating hydrometer.

designing the hydrometer so that varying quantities of liquids displaced show as large variations in floating depths and by calibrating these floating levels, the hydrometer provides an easy way of measuring relative density; and so this is the commonly used method of checking milk purity, and alcohol content of spirits.

(b) Force

When energy is expended to influence the behaviour of an object, then force is said to be present and the effect of a force or a number of forces is seen and can be measured in the form of movement. If when ploughing the plough fails to follow a true course behind the tractor, it is possible to visualise the forces acting on the working plough body and intelligent adjustments can solve the problem.

All objects around us are continually subject to two common forces acting upon them, gravity and air pressure, and these two examples will serve to illustrate some of the important features of forces.

Gravity

All objects and materials are attracted towards the centre of the earth by this force; like all motivating forces, gravity has direction. Whatever the size or shape of an object, gravity acts on it as a whole; and from the behaviour of an object in response to the force of gravity, there seems a point to which gravitational force is focused—termed the centre of gravity. Any object will respond to the force of gravity; that is, in most cases it will fall or change its position unless adequately supported to prevent this and remain stable. A tractor on a loaded trailer is stable if the centre of gravity is vertically over the supporting base area provided by the wheel-base. With most tractors, the centre of gravity lies within the transmission housing, level with the front of the tyres on the rear wheels. Being able to visualise the centre of gravity position of his tractor, gives the operator the ability to assess the tractor's stability on sloping land or when driving up a silage clamp. In the same way, a two-wheeled trailer, having a triangular supporting base formed by the two wheels and drawbar hitch, can be loaded intelligently to keep the centre of gravity low and over the widest part of the supporting base.

The force of gravity, like all other forces, has magnitude; and in speaking of the weight of an object we are in fact measuring the amount of the force of gravity acting on it; and so weight units are often used to describe the size of a force, e.g. kilogrammes force (kgf) but the more correct unit of force to use is the newton (N). Unusually for the S.I. system, there is no tidy relationship between these two units, and 1 kgf = 9.8 N.

Air pressure

On the surface of the earth, we exist at the bottom of an 'ocean' of air several kilometres deep. This depth of air produces a pressure of about one kilogramme on each square centimetre on all surfaces. Here is an example of a force not being apparently focused in one place but acting evenly over an area and producing a pressure which is measured as force per unit area. This force has a direction, at right-angles to the surface on which it is acting. Except in the case of a vacuum-filled container, this air pressure acts on the opposite sides of surfaces as well and equal and opposite forces produce no reaction.

Effects of forces

All objects are subject to forces acting on them all the time; and if an object does not change its motion, that is it remains stationary or continues movement at constant speed in constant direction, then each force acting on the object is counteracted by an equal force acting in an opposite direction. For example, a vehicle standing on a river bridge is being pulled down by the force of gravity and at the same time the bridge is exerting an upward force to equalise gravity. In the same way, a cricket ball, if it were not subject to air resistance and gravity, would continue on a straight path out into space when thrown.

It follows that if forces acting on an object are not equalised or opposite, then the object reacts by a change of motion and direction. These changes can be determined diagrammatically by

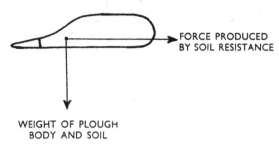

FORCE PRODUCED
BY SOIL RESISTANCE

WEIGHT OF PLOUGH
BODY AND SOIL

Fig. 1.3   Forces acting on a plough body.

drawing the *parallelogram of forces*. The parallelogram is drawn with two of its sides having lengths relative to the magnitude of the forces, and drawn at the same angles as those forces are acting on the object. The resulting reaction of the object will be related to the diagonal of the parallelogram in magnitude and direction. Visualise this parallelogram of forces applied to a plough body in work: there is resistance to its forward movement and its weight acts downwards. In what direction must it be pulled?

Friction

Whenever any object is resting on a supporting surface its weight is counteracted by an equal force exerted upwards by the supporting surface. The two surfaces are therefore being pushed together by pressure and grip between them occurs. When a force is used to try to slide the object along the surface, this grip provides a resisting force known as *friction*. Its magnitude can be determined by finding the force necessary to overcome it. Its size is influenced by the roughness of the two surfaces; and even in the case of two polished steel surfaces, there are microscopic pits and projections which are pushed together by weight and support forces to produce a grip between the surfaces. But the compressing forces have an important influence on friction. An important fact to realise is that friction is independent of the area of the surfaces in contact. On similar surfaces, weight is the only factor influencing the amount of friction because if surface areas in contact are reduced then the pressure per unit area is naturally increased. If an experiment is performed to verify this statement, it will be noticed that the force necessary to start an object moving is greater than that necessary to keep it moving. Two types of friction are recognised, *limiting friction* resisting movement when stationary and *sliding friction* producing resistance while in motion. If the microscopic character of two surfaces in contact can be visualised, limiting friction is present when a certain amount of interlocking between pits and projections occurs, while when sliding on the projections of each surface sliding friction is produced (fig. 1.4).

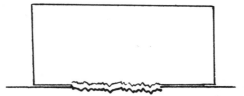

Fig. 1.4 Microscopic view of surfaces in contact between which there is friction.

Anyone attempting to move some heavy object would usually place rollers under it or suspend it on wheels; and, at first sight, it looks as though separating the two surfaces in contact provides a solution to the problem of friction. But while sliding has been eliminated, the weight of the object has now been concentrated on a much smaller area and this has the effect of indenting the surfaces at points of contact. The rollers or wheels, in order to move, now have to climb out of the indentations or produce moving indentations. There is therefore resistance to movement still present, known as rolling friction. The degree of surface indentation, which produces this rolling friction, is influenced by the weight carried and the hardness of the surfaces.

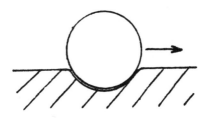

Fig. 1.5 Indentation of a surface bearing a rolling weight, causing rolling friction.

The rolling resistance or friction of the tractor front wheel on a soft seed bed is often considerable and is usefully overcome when all wheels are powered; under soft conditions the overcoming of rolling friction by four-wheel drive may be more important than the extra grip achieved.

Work

When any farm operation has to be carried out force has to be exerted. But nothing is achieved unless the force being exerted is effective in producing movement: when this occurs 'work' has been done. The word 'work' used here is not the activity for which you get paid but has a scientific meaning: it may be defined as a force acting over a distance and is measured in force-distance units, newton-metres (Nm). If a bag of feed has to be lifted 1.5 metres on to a trailer and a gravity pull of 240 newtons on the bag has to be overcome to do it, then 360 newton-metres of work has been done:

$$240 \text{ N} \times 1.5 \text{ m} = 360 \text{ Nm}$$

Compound units are inconvenient and complicated and so newton-metres are called joules:

$$1 \text{ N} \times 1 \text{ m} = 1 \text{ Nm} = 1 \text{ J}$$

An amount of work, measured in joules, can be made up in various ways. In the bag-lifting example above, a half-weight bag lifted through twice the height would have resulted in the same amount of work being carried out.

Forces: change of direction

When a sack of corn has to be lifted to 2 m high, a man may pick it up from the ground and with a throw and push may, with difficulty, deposit the sack at the height necessary. He can provide the necessary force, but it is awkward to perform this task easily. If this force could be redirected to lift the sack all the way, then the job would be much easier. A length of rope and a pulley attached high up will achieve this result; by attaching the load to one end of the rope, passing the rope over the pulley and pulling with effort on the other rope end, the sack can easily be lifted to the desired position. This is an example of a simple machine which permits the performance of work more conveniently.

Change of magnitude

When confronted with a weight which is impossible to lift manually, the weight can often be moved if some form of leverage can be used. A farm trailer with a punctured tyre can be lifted using a long bar bearing on a pillar of bricks or a block of wood. How much leverage is achieved? The leverage is obviously connected with the length of bar between pivot and load, and the length between pivot and effort. Here is another simple machine which produces changes in the size of forces.

(c) Simple machines

A simple machine permits the performance of work more conveniently. Whenever a simple machine is used, the force put into it is the *effort* and the force produced is the *load*, and so when in use:

Work Put In   = Resistance of Machine +   Work Produced
(Distance × Effort)   (Friction)   (Distance × Load)

A simple machine can increase or decrease the forces involved or it can increase or decrease the distances moved.

The force increase using a machine is known as its *mechanical advantage* and is numerically expressed as:

$$\frac{\text{Load}}{\text{Effort}}$$

The distances moved are stated as the *velocity ratio* of the machine, which is found from the expression:

$$\frac{\text{Distance Moved by Effort}}{\text{Distance Moved by Load}}$$

For a frictionless machine these two expressions would be equal, but such is never the case and the performance of a machine is known as its efficiency expressed as a percentage:

$$\text{Efficiency} = \frac{\text{Mechanical Advantage}}{\text{Velocity Ratio}} \times 100$$

Pulley systems

An arrangement of ropes and pulleys is an example of a simple machine used to lift heavy loads: remember that passing a rope around a pulley merely redirects the force conducted by the rope. By supporting a load on a number of ropes, the total supporting force is shared between all the supporting ropes. In fig. 1.6 a simple arrangement is shown; in practice the pulleys at the top are carried on a common axle and there is the same arrangement in the lifting block.

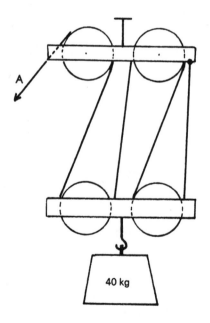

Fig. 1.6   Pulley system.

Estimate the mechanical advantage of this pulley system (disregarding friction, which in practice is an important limitation of efficiency) by calculating the load on each rope and remembering that the rope at A is the last supporting rope redirected over a pulley. In the same way, estimate the velocity ratio by calculating how much rope must be pulled at A to shorten all the supporting ropes by 100 mm each.

Lever systems

One of the commonest improvisations to move a heavy load is to use a long rigid bar as a lever and, by pivoting on something firm, use it to magnify the force that can be manually exerted. There are three ways of using a lever in which the relationships of the effort (E), the load (L) and the pivot vary; the classes of levers are as follows:

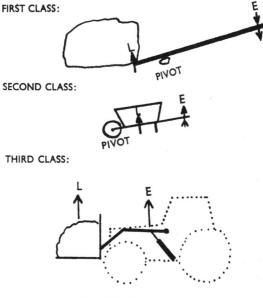

Fig. 1.7 Lever systems.

When using a bar as a lever, a firm support on which to pivot is an essential and the amount of leverage achieved depends on the relative distances of effort and load from this pivot.

Experimentation will soon prove that, using a lever having little or no friction at the pivot:

Effort × Distance from Pivot = Load × Distance from Pivot

We have assumed that the forces of load and effort have been acting at right angles to the lever, but this is not always so. How will other directions of force effect the leverage? In answering this it must be emphasised that the lever is merely a transmitter of force and the important features are the directions of the forces relative to the fixed point, the pivot.

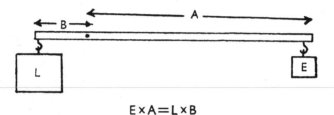

E × A = L × B

Fig. 1.8 Leverage principle.

The basic principle of levers may now be restated to answer the above question:

Effort × Perpendicular Distance from Line of Force to Pivot
=
Load × Perpendicular Distance from Line of Force to Pivot

This statement now provides for forces acting on a pivoted system from any direction. When a force produces rotation it is known as torque and is measured in force/distance units such as kg m. Turning force (or torque) transmitted by shafts is stated in these units.

Investigate the hydraulic lift arrangement of a tractor and calculate the torque on the cross-shaft, from which the lifting force seems to originate, for a given implement weight.

Obviously any object able to pivot can rotate in either of two directions and forces may produce conflicting torques, as will be seen in fig. 1.9, where the load is producing an anti-clockwise torque and the effort is counteracting with a clockwise torque.

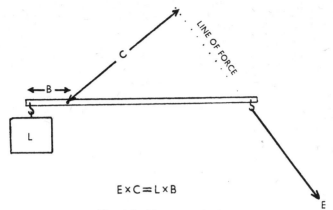

$$E \times C = L \times B$$

Fig. 1.9 Torque principle.

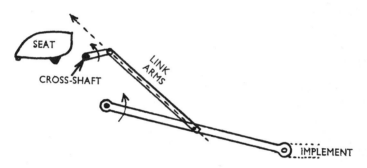

Fig. 1.10 Tractor hydraulic lift.

The principle of torque is stated as: equilibrium will only exist when the anti-clockwise torque equals the clockwise torque about a pivot. In the case of the tractor hydraulic lift, the implement load produces a clockwise torque on the cross-shaft and to lift the implement this must be more than equalled by an anti-clockwise torque produced by the internal hydraulic ram.

Wheel and axle

When a force has to be converted to a rotary motion, a crank is often used as in the case of a bicycle pedal or a tractor starting handle. But only in one position is the maximum torque produced, namely when the crank is at right-angles to the direction of the force. This disadvantage may be overcome if the crank is replaced by a wheel and the force is constantly applied to a point on the circumference of the wheel as, for instance, the steering wheel of a tractor. A wheel is really a continuously acting lever in which the distance by which the force is multiplied to give the torque is always constant, namely, the radius of the wheel.

To find the mechanical advantage of this simple machine, the simple lever may be imagined and the simple leverage principle applies (fig. 1.8):

$$\text{Mechanical Advantage} = \frac{R}{r}$$

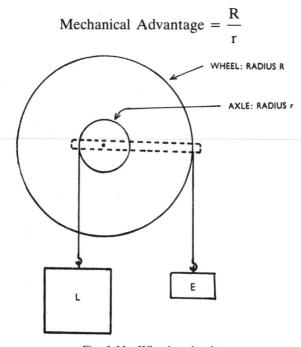

Fig. 1.11   Wheel and axle.

In considering the velocity ratio of a simple machine where rotation occurs, it will be seen that when one revolution has occurred the rope supporting the load will be shortened by one circumference of the axle while the effort rope will be lengthened by one circumference of the wheel

$$\text{Velocity Ratio} = \frac{2\pi R}{2\pi r} = \frac{R}{r}$$

(d) Transmission of power

In the transmission system of a tractor, or the mechanism drive of a seed-drill, force and movement are transmitted from shaft to shaft by means of wheels and changes in force and speed are often required. These are achieved by combining many wheel and axle systems. The relative sizes of the wheels in action gives the clue to the effects produced. First consider two wheels of different sizes in frictional contact, where the larger is the driver and the smaller the driven.

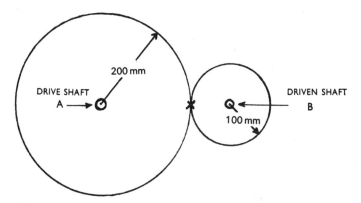

Fig. 1.12 Transmission of power.

In fig. 1.12 the drive shaft (A) has a torque of 400 kg mm. What is the torque of the driven shaft (B)? If these two wheels were two levers the ends of which touched at x, the following statements could be made:

$$\text{Torque on A} = 400 \text{ kg mm}$$
$$\text{Distance to x} = 200 \text{ mm}$$
$$\text{Force at x} = \frac{400}{200} = 2 \text{ kg}$$

If this force is transferred to the small wheel with no loss:

$$\text{Force at x} = 2 \text{ kg}$$
$$\text{Distance to B} = 100 \text{ mm}$$
$$\text{Torque at B} = 100 \times 2 = 200 \text{ kg mm}$$

When power is transmitted by a large wheel to a smaller wheel, power decreases in proportion to the relative sizes of the wheels in action.

A change in speed also occurs; trace the amount of the circumference of each wheel which passes the point x when the large wheel revolves once. In one revolution of A:

One complete circumference passes x (i.e. $2\pi r$)

$$= 2 \times \frac{22}{7} \times 200 = 1{,}260 \text{ mm}$$

If the edge of the small wheel passes x by the same amount, the shaft B and wheel revolve:

Circumference of small wheel $= 2 \times \dfrac{22}{7} \times 100 = 630 \text{ mm}$

Therefore, the wheel and shaft B revolve $\dfrac{1{,}260}{630} = 2$

When movement is transmitted by a large wheel to a smaller wheel, speed is increased in proportion to the relative sizes of the wheels in action.

Any one size character of the wheels may be compared: their radii, diameters, or circumferences.

In most systems, frictional contact between wheels is not good enough and so the edges of the wheels are cut into teeth which mesh together and produce a positive drive. If the wheels are not in contact but some distance apart, the drive is transmitted from one to the other by a belt (frictional drive) or by a chain with links to fit the teeth of the wheels which are known as sprockets, producing a positive drive. In some systems driver and driven gear-wheels are linked by a chain of gear-wheels, but so long as drive is transmitted through the edges of such wheels, they are known as idlers and they do not influence the changes of speed or power; each idler in a chain of gears does, however, reverse the direction of rotation.

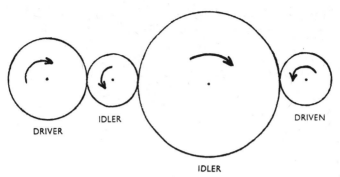

Fig. 1.13 Train of gears.

Since when two gear-wheels mesh the teeth have to be mathematically accurate and identical, their numbers are often used as an expression of the sizes of the wheels.

The velocity ratio may easily be calculated:

$$\text{Velocity Ratio} = \frac{\text{Number of Teeth of Driven Wheel}}{\text{Number of Teeth of Driving Wheel}}$$

and this figure is the number of revolutions of the driving wheel or shaft to one revolution of the driven wheel or shaft.

In the gearbox of a tractor several sets of gears may be combined to transmit the drive to the back axle. To calculate the overall effect, the ratios are multiplied together, the effect of each pair of meshing gears written as a fraction with the number of teeth of the driven wheel above the number of teeth of the driving wheel. When the expression is worked out, the numerical answer indicates how much greater or lesser the speed of the engine (the effort) is compared with the backaxle (the load).

Inclined plane

The last example of a simple machine that needs to be understood is the inclined plane, although at first sight the device does not look like a machine because it has no obvious 'mechanism'. But it does come within the definition of a simple machine by reducing the amount of force necessary to perform a task.

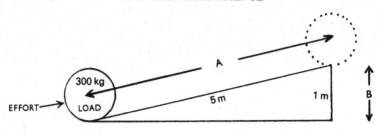

Fig. 1.14   Inclined plane.

Imagine you have to lift a 182-litre barrel of molasses weighing over 250 kg on to a stand so that it may be tapped and run into a bucket. The easiest way to do this would be to roll the barrel up a plank (an inclined plane) to the required height and position. What effort would be needed?

In fig. 1.14 a slope has been constructed with a gradient of 1 in 5, i.e. one metre increase in height for 5 m travelled up the slope. Since we know the distances involved, what is the velocity ratio of the arrangement?

$$\text{Velocity Ratio} = \frac{\text{Distance Travelled by Effort}}{\text{Distance Travelled by Load}}$$

$$= \frac{5}{1} = 5$$

It may have been noticed in previous examples that, where friction is disregarded, the mechanical advantage and velocity ratio are equal. Assuming this to be the case here, then the mechanical advantage = 5, but

$$\text{Mechanical Advantage} = \frac{\text{Load}}{\text{Effort}}$$

$$\therefore \qquad 5 = \frac{300}{?}$$

$$\therefore \text{Effort} = \frac{300}{5} = 60 \text{ kg}$$

There is always a relationship between effort and load linked to the gradient of the slope, e.g. the gradient is 1 in 5 and the effort is $\frac{1}{5}$ of the load; if the gradient was 1 in 200 the effort would be $\frac{1}{200}$ of the load.

A screw thread and a worm gear are examples of inclined planes 'wrapped around' shafts. The measurements from which the velocity ratio is calculated are shown in fig. 1.15 and it will be obvious that:

Distance moved by Effort(e) = Circumference of Shaft, and,
Distance moved by Load (l) = Pitch of the Thread.

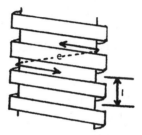

Fig. 1.15 Screw thread.

But, in practice the friction of such a device is considerable and the mechanical advantage is very different to the velocity ratio. This friction is an advantage because it means that in the case of a screw jack, the load cannot cause the jack to screw itself down; also, in the case of a worm gear drive—as used in the steering-box of a vehicle—the drive cannot be reversed in direction and the lash of the front wheels is prevented from turning the steering wheel.

(e) Hydraulics

In the examples studied so far, work has been seen as force passed by some rigid object such as a bar or a wheel and moving a certain distance usually in a straight line. Force need not necessarily be focused as energy on a single path, but might be distributed over an area. If the force supporting a seated person were limited to one spot, the person would not sit for long.

When physical energy is exerted over an area, it is known as pressure per unit area. The air pressure which all objects on the earth's surface have to withstand is a good example. The earth's surface is the bed of an 'ocean' of air some thirty miles deep, and the pressure produced by the air above amounts to 1,033 g/cm². Fortunately the inside as well as the outside of objects is at the same pressure, resulting in equality except when a vacuum is formed and the equalizing pressure inside the container is removed.

If the total pressure over a large area needs to be stated, it is known as thrust, measured in force units. The air pressure on an exercise-book page amounts to a thrust of more than 300 kg.

$$\text{Thrust} = \text{Pressure} \times \text{Area}$$

The simple machine which makes use of energy transfer as pressure is the hydraulic press, but to understand its working one or two facts concerning fluids must be stated.

A solid substance has characters such as hardness and rigidity because the molecules of which it is composed are closely packed and rigidly attached to each other so that the substance can neither be compressed nor changed in shape. Examples easily come to mind, such as rubber and sheet metal, where this generalisation about molecules in solids is not strictly true. But contrast the generalisation about solids with the case of a fluid substance. In this case the molecules are free to move relative to one another and the substance flows to fill and shape itself to any container in which it is placed. With most substances an increase in heat energy brings about a change in state from solid to fluid. If, in the fluid state, the molecules of the substance are tightly packed together, it is a liquid and is not, for all practical purposes, compressible. If the molecules have become so active and loosely packed as to be entirely unattached to each other, then the substance is a gas and the molecules distribute themselves equally throughout the space in which they are contained. In hydraulics, the two important facts to note about liquids are:

1. They shape themselves to the container
2. They are incompressible

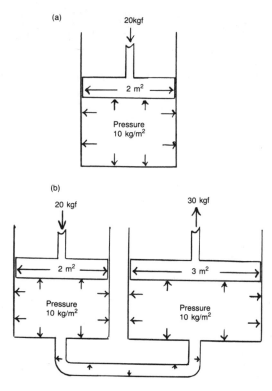

Fig. 1.16 Hydraulic ram.

Because a liquid cannot be compressed, it distributes pressure applied to it equally in all directions. In fig. 1.16 (a) a cylindrical container filled with liquid is closed by a piston of 2 m^2 surface area and the thrust on this piston is 20 kgf.

The piston is applying a pressure of 10 kg per m^2 to the liquid which, since it is fluid and incompressible, is distributed equally to all containing surfaces. In fig. 1.16 (b) another cylinder has been connected to the first, the liquid in it being at 10 kg per m^2 pressure and acting on a piston of 3 m^2 area.

The thrust on the large piston is 30 kgf magnified by the hydraulic press in proportion to the relative sizes of the two pistons. In a tractor hydraulic system, oil under pressure—pressures of thousands of kilos per square metre, maintained by a pump—is used to activate pistons (rams); and, following from the illustrated example, the desired force or thrust is achieved by using a piston of certain area.

Power

So far energy has been measured as a physical action labelled work, being a force measured in kgf or N, acting through a distance usually in metres. If about a tonne of manure with a gravity pull of 10,000 N has been forked from the floor of a loose-box on to a trailer 1.5 m high, then 10,000 N × 1.5 m = 15,000 J of work has been done.

If two men have been working at this job side by side, and one has done his share of the work quicker than the other, they may have done the same amount of work when they have finished but while they were forking the manure the faster man was moving more manure in a period of time than the slower man. As soon as the amount of work in a certain time, or rate of work, is considered, this is measured as work and the time dimension is added to the work units. If it takes the slower of these two men longer to produce the same amount of work, he has not expended as much energy as the faster worker. If the slow man took 2,000 seconds, he produced

$$\frac{15,000 \text{ J}}{2,000 \text{ s}} = 7.5 \text{ J/s}$$

while the faster worker took 1,000 seconds, the power he produced was

$$\frac{15,000 \text{ J}}{1,000 \text{ s}} = 15 \text{ J/s}$$

The rate at which work is done indicates the power involved and is measured in watts:

$$1 \text{ joule per 1 second} = 1 \text{ watt}$$

In the imperial measurement system, the unit of power was the horsepower and 1 horsepower is 746 watts.

The scientific measurement of power therefore includes three components: force, distance and rate (time).

It is an important physical law that energy is not spontaneously produced nor does it disappear. Energy is neither created nor destroyed but is converted from one form to another. Many of the devices used on the farm are energy convertors.

In further chapters, ways in which these other forms of energy are used on the farm will be discussed.

(B) Electricity

An important source of energy on the farm, electrical power from the mains supply provides the driving force of electric motors used in 'fixed' positions to drive milking machine vacuum pumps and barn machinery. Electricity supplied by batteries on tractors relieves us of the effort of starting the engine by manually 'swinging the starting handle'. Before considering devices using, producing, or changing electricity, we must try to answer the question: what is electricity? And discover how it behaves.

Atomic structure of matter

Every substance consists of atoms of elements, originally defined as the smallest particle of a substance which could exist, before scientists started atom smashing. Each atom consists of a nucleus made up of two types of particles, neutrons and protons; orbiting around the nucleus like satellites is a third type of particle, called electrons. A neutron is electrically neutral, while the other types of particles are electrically charged—the protons being positively charged while the electrons are negatively charged. All atoms consist of these basic particles and atoms of different elements differ in the numbers of these particles present.

In the normal atom of any element, the number of protons (positive charges) equals the number of electrons (negative charges) and the atom is electrically neutral. But atoms of different elements vary in the tenacity by which the electrons are held, and electrons may be easily removed from atoms of some elements such as copper. If this happens, the atom becomes short of a negative charge and exerts an attraction on the electron of a neighbouring atom. In this way a chain reaction of exchanging electrons occurs and this causes a flow of energy, electrical energy. Remember how small atoms and electrons really are: if in one second about $6\frac{1}{4}$ million million million electrons pass one point on a wire then an electrical current of 1 ampere is flowing.

If the atoms of an element readily exchange electrons, then that element is a conductor; while if electron exchange cannot take place, the element is an insulator. The conductivity of any substance is a characteristic which varies with the electrical force involved: for instance, air is not normally a conductor, until sufficient electrical force causes a spark through air. To produce a spark across an air gap of 25 mm requires an electrical force of 78,000 volts.

(a) Electrical circuits, power and supplies

Circuits

Electricity as electron flow cannot move along an isolated length of a conductor but flows in a circuit from a device, such as a battery or generator, and back again. The device such as a battery produces an attraction for electrons at one electrode and a supply of electrons at the other, that is, there is a force between them measured in volts: a measure of electrical pressure providing available energy. The mains supply usually has a pressure of 240 volts between the two wires, while a tractor battery has a pressure of 12 volts.

All materials, however good as electrical conductors, have some resistance to electron exchange, and this resistance acts against the electrical pressure and is measured in ohms. The resistance of any substance usually increases with temperature rise and is also varied in a conductor by its thickness.

In spite of the resistance in a circuit the voltage forces some electrical energy along the circuit and the quantity or current is measured in amperes (amps).

There is a constant relationship between volts, ohms, and amps, expressed by Ohm's Law:

$$\text{amps} = \frac{\text{volts}}{\text{ohms}}$$

An appliance with a resistance of 60 ohms, when plugged into the mains supply of 240 volts, allows 4 amps of current to flow. In such an appliance, the effect of voltage overcoming resistance causes some of the electrical energy to be converted into heat energy and, made of the right substance and thickness, the conductor may glow red hot and radiate heat or glow white hot and produce light.

Electrical power

The power available in a circuit which can be used for mechanical energy or heat energy is measured in watts, and appliances using electrical power are rated in watts or, in the case of large appliances, in kilowatts (= 1,000 watts). The wattage available in a circuit is mathematically found by multiplying volts and amps. The maximum power available from the domestic mains is limited by the amount of current which can flow along the wiring, usually about 30 amps. The available power is therefore:

$$30 \text{ amps} \times 240 \text{ volts} = 7,200 \text{ watts}$$

From the expression watts = volts × amps, the circuit loading for any particular appliance can be calculated. If an electric fire rated at 1 kilowatt is plugged into the mains, how much current will flow in the circuit?

$$1,000 \text{ watts} = 240 \text{ volts} \times ? \text{ amps.}$$

$$\frac{1,000}{240} = 4 \text{ amps. (approx.)}$$

In mechanical terms, power is the rate of doing work; and similarly, in measuring electrical power time is a consideration. The commercial unit by which electrical power is sold is the kilowatt-hour (kWh) or unit. One unit of electricity has been used when 1,000 watts have been consumed for one hour. This may take the form of a one-bar electric fire (1,000 W) burning for 1 hour, a 100 W electric light bulb burning for 10 hours, or 1 h.p. electric motor (880 W) running for about $1\frac{1}{4}$ hours.

Measuring consumption

The electricity supply to a house or farm always passes through a meter, usually situated near the fuseboard and main switch. Consumption of electricity can easily be checked by reading the meter at intervals, and the difference between readings is the amount consumed. To read an electricity meter inspect the dials and, starting with that recording 10,000 kWh, write down the last figure the needle has passed and do this for each dial—taking care because the dial needles revolve in different directions—and disregard the last dial (usually red) which measures $\frac{1}{10}$ kWh.

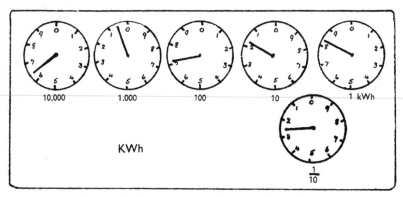

Fig. 1.17   Electric meter dials (reading:60718).

Electricity supplies

Since electrical power (watts) is derived from both electrical pressure (volts) and current flow (amps), it is obvious that power can be provided either from low voltage supply and a heavy flow of current or high voltage and small current:

480 watts = 240 volts × 2 amps = 12 volts × 40 amps

Both these ways of providing electrical power have disadvantages. In the first case, high voltage is a safety hazard to operators and needs good insulation of all wiring to prevent leakage and, in addition, the contacts of switches become burnt because of sparking when contacts open. In the second case, the heavy flow of current needs thick wires to carry the amps without too much resistance. This is why the cables connected to a tractor battery are so much thicker than those carrying mains electricity. When electricity has to be carried over distances, then a loss of energy occurs and this energy loss is proportional to the current^2; so high currents mean high losses over distance. To distribute mains electricity from generating stations through grid lines, very high voltages of 30,000 volts and more are used to cut down losses and permit the use of relatively light cables to carry the load. To overcome the need for thick insulation, the wires are carried on pylons overhead.

Forms of current

Earlier it was stated that electricity could only flow around a circuit from a point of high pressure to a point of low pressure. But, by experimenting with any battery powered device, such as a bulb or a small electric motor, it is found that the device works whichever direction the current is flowing; it is the electrical energy flowing past a particular point in the circuit which activates the device.

Two forms of current are commonly encountered, direct current (D.C.) and alternating current (A.C.). In a circuit connected to a battery, current flows in one direction from the positive terminal to the negative terminal and, if measured with a voltmeter, would show a constant voltage. This is *direct current*. Direct current is the form of current used in tractors and cars where a battery is an essential part of the system.

To supply mains power, direct current has the disadvantage that its voltage cannot be easily stepped up or down. Therefore mains supplies are almost always in the form of alternating current which can be easily transformed in voltage. In a circuit carrying A.C. current the direction of flow is continually reversing rapidly.

The difference between D.C. and A.C. may be shown graphically, where X and Y are the two wires from the electricity source (see Fig. 1.18).
Alternating current changes through the cycle shown in $\frac{1}{50}$ second, and is therefore said to alternate at 50 cycles per second. From the graph it will be seen that there is no voltage difference between the two wires X and Y twice in each cycle so the power is really off 100 times a second, but the alternation is so rapid that electrical

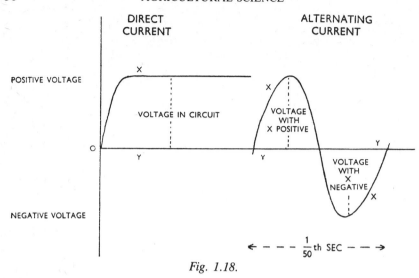

Fig. 1.18.

devices are unaffected. The two wires of a supply of direct current are normally labelled *positive* and *negative*, but in the case of a supply of alternating current the polarity changes result in the two wires being labelled *line* or *live* and *neutral*.

3-phase current

The type of alternating current so far discussed is known as single phase supply, being power carried on two wires or conductors. There is, however, a limit to the amount of energy available in a single phase supply, and if a farmer requires motors of several horsepower to drive barn machinery then more electrical energy is needed. A heavy drain on a single phase supply could produce a dangerously high current in the supply wiring and affect the supply to other users.

The commonsense answer to such a problem would seem to be to provide other extra supplies to the farm to meet the need. This would mean several pairs of wires might be used and, if three such pairs were used, three 'quantities' of electrical energy would be available. If it is remembered that electrical energy is derived from a voltage or pressure between two wires or conductors, then it is obviously possible for these three 'quantities' of electrical energy to share conductors in such a way that only three

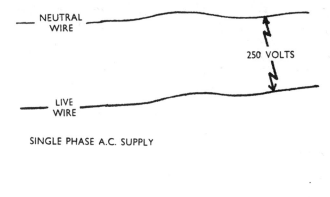

NEUTRAL
WIRE

250 VOLTS

LIVE
WIRE

SINGLE PHASE A.C. SUPPLY

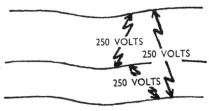

250 VOLTS

250 VOLTS

250 VOLTS

Fig. 1.19 Three phase A.C. supply.

conductors are necessary; there existing a pressure of 240 volts between any two of them. The alternations of these three 'quantities' of electricity are out of step and this form of supply is therefore known as 3-phase supply.

It is important not to confuse the three lines of a three-phase supply with the 'live', 'neutral' and 'earth' conductors of a single phase supply. It is possible for small loaded appliances to use only

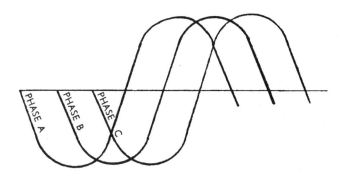

PHASE A PHASE B PHASE C

Fig. 1.20 The alternations of 3-phase current.

one phase of a three phase supply, but when connections are being made, the connecting of a number of appliances should be so distributed over the three phases that the phases are as equally loaded as possible. From this discussion, it should be obvious that such connections should never be attempted by the do-it-yourself enthusiast but, like all electrical works, require the services of a highly-trained expert.

Conductors and insulation

If electrical energy is to pass through a substance, as has been discussed earlier, then the exchange of electrons from atom to atom must occur. The ease with which this occurs varies with different materials. The metals, copper in particular, carry electricity easily with very little resistance and are classified as *conductors*: while materials such as asbestos, porcelain, plastics, and rubber, will not conduct electricity and can therefore be used as protective layers around electrical conductors; these are classified as *insulators*. With some materials, such as air, classification in the above categories is not possible because, while resistance is sufficient to prevent conduction of electricity at low voltages, when high voltages are present the resistance of air is insufficient to prevent a spark occurring as in an engine or when lightning occurs.

Water is a good conductor and the entry of moisture into electrical appliances, or into cables if the insulation deteriorates, is a major hazard especially on the farm. The danger to operators can be removed by ensuring that all equipment is efficiently earthed (through the large pin in a 3-pin plug) so that any leakage travels to earth by an easier route than through anyone who touches the faulty equipment.

(b) Electrical effects, movement and chemical change

When a flow of electricity occurs through a conductor—which may be solid, liquid or gas—several effects may occur.

When current passes through a wire, magnetic forces are created which can cause components made of ferrous metals to move, and devises using this movement will be studied.

Any conductor, in which resistance has to be overcome for electricity to flow, often becomes heated, some of the electrical energy appearing as heat energy. When electricity is passed through solutions of some compounds chemical reactions take place.

First consider the simplest of these effects of agricultural importance, the heating effect of an electric current.

Fuses

When any electrical circuit is installed, the probable load and current can be estimated and the wiring used is sufficient for the load expected. Over the course of years the insulation of the wiring may be damaged or perish and cause a 'short circuit', that is the flow of current does not pass under control through the circuit, but returns to earth too easily and large currents flow. This can also happen when an appliance is faulty and short-circuits. The flow of heavy currents through overloaded wiring can produce heat and eventually a fire may break out.

Such a catastrophe can be avoided quite simply by purposely including a weak spot in the circuit where no harm can be done. This is the function of a fuse. A link of special wire which, when overloaded, gets hot, melts and breaks the circuit. For a fuse to be efficient it must consist of wire of the correct thickness and length, and these points should be remembered when replacing a fuse. A repeatedly burnt out fuse is not a sign that thicker fuse wire is indicated but that the circuit is overloaded and the wiring and appliances being used should be inspected.

Heat from electricity is needed to heat up heaters in order to warm air on grain driers and to light lamps. In all cases the hot wire consists of material which will not burn or melt and the appliance is safely constructed. The loading of heaters and lamps is measured in watts or kilowatts; heaters of 36 kilowatts and lamps of 100 watts being common.

Magnetic force producing movement

The magnetic properties of a bar magnet are well known. The magnetism is strongest at the ends of the magnet and these are

known as the poles; when the magnet is free to swing, the end
pointing to the earth's north is known as the north pole of the
magnet. If the magnetism is studied further, iron filings sprinkled
on a sheet of paper above the magnet will indicate that a magnetic
field consisting of lines of force linking the poles exists around the
magnet.

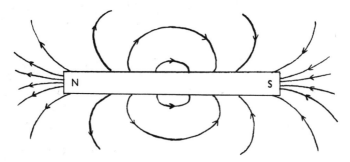

Fig. 1.21   The magnetic field around a bar magnet.

If the magnetic influence of a wire through which electricity is
passing is investigated with a compass or iron filings, the lines of
force of the field neither go to nor from the wire but are dispersed
as concentric rings around the wire.

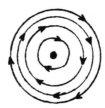

Fig. 1.22   The magnetic field around an electrified wire.

Another important difference between bar magnetism and electro-
magnetism is the fact that while the magnetic field around a bar
magnet is reasonably static and permanent, the magnetic field of a
wire 'grows' from the wire when the current is switched on, is
maintained while current flows, and contracts 'into' the wire when
the current is switched off. The size and strength of the magnetic
field is proportional to the current flowing.

The magnetic force in the field surrounding a single wire is relatively insignificant, but becomes useful when the wire is wound into a coil and fields of adjacent turns of the coil combine.

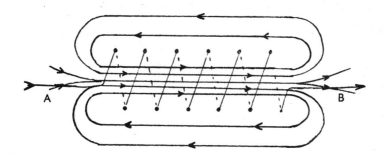

Fig. 1.23 The magnetic field around a coil.

If the collective magnetic field of the coil of wire is now investigated, the coil seems to behave like a hollow magnet and its polarity depends on the direction of current flow. The north pole end of the coil is the end at which the current is flowing anti-clockwise, easily remembered by the symbol:

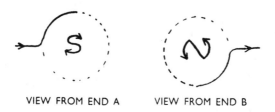

VIEW FROM END A   VIEW FROM END B

Fig. 1.24 The magnetic field around a coil (sectioned).

From the diagram showing the magnetic field of a coil, it will be seen that the lines of force inside the coil are very close together and the magnetic force is very 'concentrated'. With a substantial current flowing, this force is sufficient to move an inserted metal bar in sympathy with the magnetic force. This principle is used in the solenoid switch.

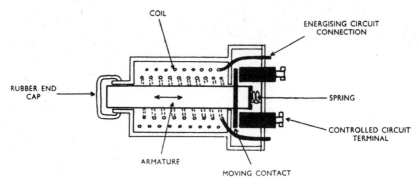

Fig. 1.25 Simplified section of a solenoid switch.

Solenoid switch

When the coil or solenoid is energised, the magnetic force produced moves the armature against the force of the spring and the contacts are thereby closed. This switch therefore permits a small energising current to close a quite independent circuit carrying a heavy current. It also has the advantage that the mechanical movement of the armature is rapid so that the contacts 'make' or 'break' rapidly with the minimum of arcing and burning. Although it is possible to move manually the armature by pressing on the rubber cap, this should be avoided as the movement is not rapid enough to prevent arcing.

In the above example, moving current induces lines of magnetic force which produce mechanical movement. The reverse of this process is easily possible. A magnet moving close to a wire induces a current in that wire and, as one might expect, the direction of magnetic movement influences direction of current flow. A current is induced when a conductor cuts a line of force in a magnetic field.

Dynamo

When a coil of wire rotates in a magnetic field, such as that present between the poles of a horseshoe magnet, an electric current is induced in the coil of wire.

On close examination, it will be seen that in one rotation one side of the coil is passing across the lines of force in one direction and then the other. This has the effect of reversing the direction of

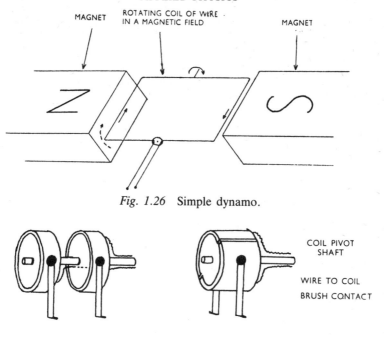

MAGNET   ROTATING COIL OF WIRE ·
          IN A MAGNETIC FIELD   MAGNET

Fig. 1.26 Simple dynamo.

COIL PIVOT
SHAFT

WIRE TO COIL

BRUSH CONTACT

SLIP RINGS   COMMUTATOR

Fig. 1.27

current flow, so in a rotating coil an alternating current is induced; the speed of alternation depending on the speed on rotation. To feed this electric current to an external circuit, connection must be made to the rotating coil by means of contacts brushing on rotating metallic rings mounted on the coil pivot shaft.

If slip rings are used, the connections between the ends of the coil and the external circuit are continuous and an alternating current flows, but if a split ring commutator is used, every half rotation reverses the connections between the rotating coil and the external circuit, so that a direct current flows around the circuit.

The generator of a vehicle which has a battery in its electrical system has to produce a direct current to enable battery charging to take place.

On lightweight motor-cycles and pedal cycles a simpler form of generator is used, in which a magnetised rotor revolves between stationary coils so that no commutator is needed; but such a device produces only alternating current and cannot charge a battery.

Electric motors

In studying the dynamo, it has been seen that mechanical movement involving magnets can produce a conversion of mechanical energy to electrical energy; the reverse is just as easily achieved. If an electro-magnet is suspended in a magnetic field and is free to rotate in that field, when energised it will swing so that unlike poles are adjacent. If, in this position, a commutator reverses the direction of current flow and therefore the polarity of the electro-magnet, it will continue to turn and continuous rotation can be produced.

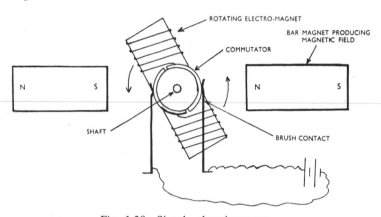

Fig. 1.28 Simple electric motor.

The magnetic field in which the armature rotates can be produced either by a permanent magnet or by field electro-magnets. In the former case, the motor will only run on direct current; but when the field is electrically induced, the motor functions equally well on alternating or direct current.

Electric motors used to drive machinery appear to have a complicated structure, but are only developments on this basic principle. A 3-phase motor is like three armature windings wound in one armature, and each phase separately activates a separate winding as the armature revolves. Using 3-phase current connected to six field coils in sequence, the fluctuation of voltage in each phase and coil in turn produces a revolving magnetic field which 'drags' a metallic unmagnetised armature into rotation without the need for a commutator. Such a motor is known as a squirrel cage motor by the construction of the rotor. Its speed of revolution is in sympathy with the alternations of the current (50 cycles) to give running speeds of 500 r.p.m., 1,000 r.p.m., and 3,000 r.p.m. when lightly loaded. This type of motor has, if producing more than 2,250 W, a very high starting current and to overcome this difficulty special starting gear is sometimes necessary.

The transformer

It has been seen that electricity can induce magnetism and moving magnets can induce electricity. these two actions are combined in the transformer. It consists of a laminated iron core around which are wound two windings of insulated wire, known as the primary and secondary windings.

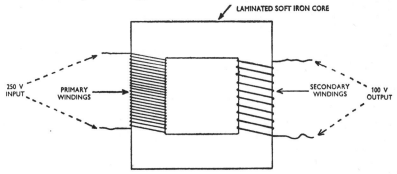

Fig. 1.29 'Step-down' transformer.

A transformer will only transform alternating current because the current induced in the secondary winding only arises when the magnetic field produced by the primary winding expands or contracts, which only happens when the primary current fluctuates. Passing electricity through a transformer does not change the wattage (amps × volts), but changes the amperage and voltage in proportion to the number of primary and secondary windings.

Twice as many secondary windings $\Big\}$ voltage × 2
      as primary windings   amps ÷ 2

Half as many secondary windings $\Big\}$ voltage × $\frac{1}{2}$
      as primary windings   amps ÷ $\frac{1}{2}$

By selecting the correct ratio of primary to secondary winding numbers any voltage or amperage can be provided by a transformer from alternating current.

'Chemical' effects of an electric current

With many inorganic compounds such as common salt (sodium chloride or NaCl) the component elements, in this case sodium and chlorine, while firmly combined in the dry state, when dissolved in water to form a solution are so loosely linked that they can separate. In a solution of salt there are separate minute units of sodium and chlorine known as ions and the separating process is known as ionisation. These components, when formed into the compound, are linked by positive and negative electrical forces and when ionisation occurs the ions exhibit these electrical charges:

$$\left.\begin{array}{c} \text{Sodium chloride} \\ \text{NaCl} \end{array}\right\} \begin{array}{c} \text{partially} \\ \text{ionises to} \end{array} \left\{\begin{array}{cc} \text{Sodium ions} \\ \text{Na}^+ \end{array} + \begin{array}{c} \text{Chloride ions} \\ \text{Cl}^- \end{array}\right.$$

By neutralising these ionic charges the ions become free and independent atoms of the component elements, this process being electrolysis and the solution in which it may be carried out an electrolyte.

Electrolysis

When an electrical circuit is completed through the two electrodes, the positively charged electrode, known as the anode, attracts negatively charged ions which, on contact become free atoms of the element. The positively charged ions are similarly attracted to the negatively charged electrode, the cathode.

When the electrolyte consists of water to which has been added a small quantity of sulphuric acid to improve its conductivity, Oxygen ions are attracted to the anode, and have their charge neutralised by acquiring a positive charge from the electrode. The uncharged atoms then appear as oxygen gas which may be collected in a tube surrounding the anode; the hydrogen ions, attracted to the cathode, behave in a similar manner. Why does

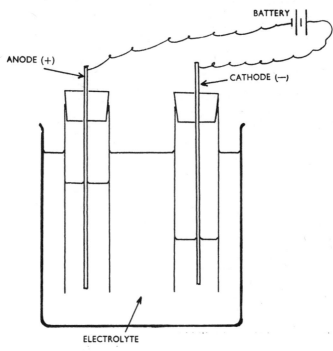

Fig. 1.30 Electrolysis.

the diagram indicate that there is more hydrogen gas than oxygen gas produced?

In the electrolysis of salt solution, the sodium released at the cathode immediately reacts with water to evolve hydrogen gas, but if a solution of a metallic compound where no such secondary reaction occurs is used, the metal becomes deposited on the electrode. This is the principle of electro-plating; metallic components being commonly plated with nickel, chromium and silver by this method.

Electrolysis is chemical change produced by electricity; the reverse process is possible and permits the chemical storage of electricity.

Simple cells

When a metal is immersed in an acid or alkali with which it reacts, it forms a compound with part of the acid, and in doing so has to acquire the necessary linking charge.

Example: Copper in sulphuric acid forms copper sulphate

$$Cu + H_2SO_4 \rightarrow Cu^{++}SO_4^{--}(+ H)$$

When a metal dissolves it gives up electrons to become positively charged. If two metallic plates, such as zinc and copper, are immersed in acid, since they dissolve at different rates they produce electrons at different rates and this difference results in a potential or pressure causing electricity to flow along a wire connecting these dissimilar metallic plates. In this case the pressure is higher at the copper electrode, so it is positive current flowing to the zinc electrode, the negative. This simple cell produces a voltage of nearly one volt, but the accumulation of hydrogen bubbles seriously limits the reaction.

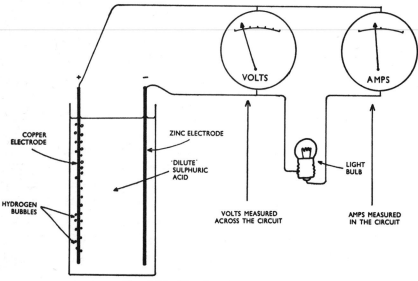

Fig. 1.31 Simple cell and circuit.

The 'dry' cell

The practical difficulties of having sulphuric acid liquid, which is corrosive, and the inhibiting effect of accumulating hydrogen gas, have been largely solved in the construction of dry cells or batteries. The zinc electrode has been converted into the outer

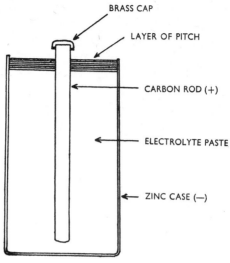

BRASS CAP

LAYER OF PITCH

CARBON ROD (+)

ELECTROLYTE PASTE

ZINC CASE (—)

Fig. 1.32 Dry cell.

metallic case of the cell; the electrolyte liquid has been replaced by a moist paste consisting of a mixture of ammonium chloride, zinc chloride, powdered carbon, and manganese dioxide, all packed around a central carbon rod which forms the positive electrode. In the interest of economy the case is only thick enough to act as an electrode for the life of the cell, and the electrolytic paste is inclined to leak through the zinc case of an exhausted cell.

The lead/acid battery

The expendable dry battery is useful where small currents are required in portable devices, but a battery which can be regenerated or charged is needed where high currents have to be provided intermittently. The lead/acid battery is a device which incorporates both the reactions we have studied; chemical change produces electricity, and this change in turn is reversed by electricity fed into the battery producing chemical change. The construction of the battery is shown in fig. 1.33, but if its working is understood then much of practical significance is straightforward commonsense.

The new battery is fitted with plates of 'spongy' or 'porous' lead, made in such a way that the whole substance of the plate

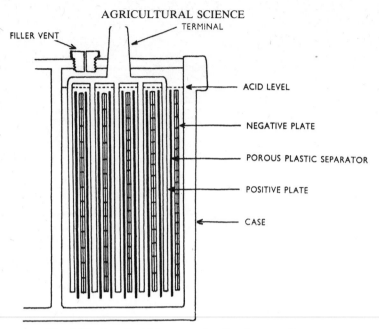

Fig. 1.33 Lead/acid battery: details of one cell.

undergoes chemical reaction and not just the surface. When a new battery is filled with sulphuric acid and first charged, the positive plates are connected to the positive direct current source and the water in the sulphuric acid electrolysed. Hydrogen is evolved at the negative plate while oxygen evolved at the positive plate reacts with the lead to form chocolate-brown lead oxide. Some oxygen is evolved and since the two gases can form an explosive mixture, dispersal of the gases should be assisted by removing the vent plugs, charging in a well ventilated place, and avoiding naked lights which might cause an explosion.

The plates are now chemically dissimilar and dissolving in acid at different rates provide a supply of direct current, rated in ampere-hours, which flows until both sets of plates have changed to lead sulphate. During this discharge the acid has become weak and been reduced in density.

Re-charging, as before, produces hydrogen at the negative plates which converts the lead sulphate into lead and sulphuric acid, and produces oxygen at the positive plates which converts the lead sulphate into lead oxide and the battery becomes fully charged. When recharging, it is important not to charge too quickly as this can cause heating and buckling of the plates which results in material dropping out of the plates and forming a short-circuiting sludge at the bottom of the cell.

If an occasion arises when a battery has to be stored, such as a combine-harvester battery being stored over the winter, it should be fully charged and this state maintained by a topping-up recharge every two months. There are two reasons for this recommendation. Firstly the electrolyte is more concentrated when the battery is charged and therefore less likely to freeze. Secondly, every time the plates change to lead sulphate and then, on charging, change back to lead or lead oxide, a small proportion of the lead sulphate cannot be changed back and the quantity of this 'fixed' lead sulphate builds up and in this way limits the life of the battery. This sulphation effect is aggravated if the plates remain as lead sulphate (uncharged) for any length of time.

To summarise the changes taking place in the battery:

	Positive Plates	Negative Plates
First charge:	Lead to Lead Oxide	Lead Unchanged
Discharge:	Lead Oxide to Lead Sulphate	Lead to Lead Sulphate
Re-charge:	Lead Sulphate to Lead Oxide	Lead Sulphate to Lead Oxide

Precautions:	Positive battery terminal connected to positive source. Charge slowly (as manufacturers' instructions). Remove vent plugs to disperse gas.
Storage:	Store fully charged.

Rechargeable 'dry' batteries (cell)

The nickel-cadmium battery, similar in size to dry batteries and used for the same purposes, such as powering torches and meters, contains chemicals which have a reversible reaction and can be recharged. They contain nickel dioxide as the positive electrode and cadmium as the negative electrode, both working in potassium hydroxide electrolyte. Electrical energy is released or absorbed by an exchange of oxygen. These cells hold their charge well, produce $1\frac{1}{3}$ volts per cell and operate well at low temperatures.

(C) Heat

(a) Heat energy and heat movement

What is heat?

Confronted by energy in other forms, we have experienced difficulties in exact definition. Heat is certainly a form of energy, but its expression as movement is not so apparent as energy in other forms. Heat energy results in molecular movement within a substance, each minute molecule having movement within itself and also relative to its neighbouring molecules. The amount of molecular energy present influences the forces of adhesion between molecules. If the amount is low, movement is slow and adjoining molecules adhere to each other giving the substance solidarity, in fact making it a solid. If heat energy is added to a mass of solid substance, its temperature rises until at a particular point the molecules cease to adhere to each other although continuing to be attracted to each other, so they may shift their relative positions and the substance can flow; it has liquefied. Should the adding of heat be continued, after a further phase in which the increasing internal energy of the liquefied substance appears as rising temperature, another critical point is reached. At this point the molecular activity becomes so great that the attractive forces between molecules become nullified and each molecule behaves independently and the substance has become a gas.

It will be seen that heat addition to substance produces two effects: temperature rise and change of state, and these will receive attention later.

Measurement of heat

There are many situations where heat needs to be measured and moved in farm materials, winter-cold soil needs to be warmed for spring growth to occur, milk has to be cooled to ensure that it keeps well and animals have to be housed at comfortable temperatures.

All objects contain heat—internal molecular energy—and since heat is contained in material, heat quantities have to be related to the amount of heat energy contained in a measured mass of material. Mechanical energy which produces force acting over a distance is measured in joules, so it is logical to measure quantities of heat energy in joules. Formerly the calorie was the unit used and 1 calorie = 4.2 joules.

In theory a mass of material could have no heat energy in it at all, in which case its temperature would be minus 273°C (absolute zero) but at normal temperatures all materials contain some heat. But they do not all contain the same amount of heat energy, even at the same temperature, because different materials react differently to amounts of heat energy. Different materials have different 'heat capacities'. The *heat capacity* of any material is the amount of heat energy needed to be added to 1 kg of the material to raise its temperature by 1 degree Celsius. Here are some examples of heat capacities:

Water	4,200 J	Ice	2,000 J
Paraffin oil	2,200 J	Copper	400 J
Dry soil	800 J	Aluminium	960 J
Grain	2,000 J	Iron	470 J

Sources of heat

Heat energy can be produced from various sources, of which electrical and chemical sources are important in agriculture. Electrical energy converted to heat energy is a convenient and easily controllable source where electricity conducted along a wire with a high resistance produces heat. Heaters are calibrated in watts, producing 1 joule of heat in one second per watt, a 2 kW heater (2,000 W) switched on for 10 minutes (10 × 60 seconds) produces:

$$2,000 \times 6,000 = 120,000 \text{ J } (= 120 \text{ kJ})$$

Chemical energy can be released as heat by burning fuels, the energy which they contain being known as their calorific value. Some examples are:

Coal   34.8 MJ/kg   Wood   18.5 MJ/kg   Fuel oil   43 MJ/kg

Organic substances fed to animals as food have an energy value:

Animal fat	39,700 J/kg D.M.
Animal protein	23,800 J/kg D.M.
Animal carbohydrate	17,400 J/kg D.M.

when digested and utilised by the animal, a large proportion of this energy appears as heat energy, so a sow being normally fed produces about 900 kJ per hour.

Heat movement

The fundamental law of heat movement is that heat moves in various ways to equalise the temperature of substances in close proximity to each other, irrespective of the actual heat energy content of those substances. The keen swimmers who enjoy breaking the ice for a morning swim leap gaily into a mass of water which contains more joules of heat than their own bodies, but this fact is little comfort when they emerge from the water blue with cold. To equalise temperatures, heat travels in three ways. It may pass through space as heat rays as, for instance, heat given off from the sun or an electric fire. This is heat movement by *radiation*, and only objects which are dense enough to obstruct the passage of heat rays will receive heat in this way.

Heat uptake is influenced by density and also by the surface character of the object. The sun's rays reaching the earth, do not heat the atmosphere but heat up the earth which in turn warms the air. The temperature of the soil is controlled by sun height and day length which cause it to increase, and night radiation and air losses which cool it. The surface character influences heat uptake because light-coloured or shiny surfaces can reflect heat; the efficiency of an electric fire is enhanced by having a brightly polished reflector. Soils of varying colours show substantial differences in heat absorption: if an area of soil is whitened by lime and its heating up is compared with adjacent untreated soil, a difference of several degrees will be noticed during a spring day. Light-coloured clothes and cars keep the occupants cooler in hot weather.

Heat movement can also occur from one object to another when they are in contact; it flows and moves by *conduction*. This often

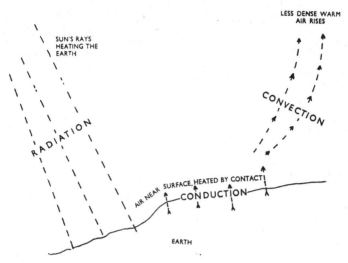

Fig. 1.34 Ways in which heat moves.

results in temperature differences between the various parts of an object or mass of substance. In very cold winter conditions the radiator of a vehicle may freeze at the bottom and very soon the water at the top of the radiator is boiling, in spite of being in contact with the ice. Different materials vary considerably in their ability to conduct heat. Generally speaking, metallic substances are good heat conductors while ceramic materials are poor conductors and are used as insulators where heat movement by conduction is undesirable.

Heat energy can move, in substances which are fluid, by currents generated within the mass of substance; a movement known as *convection*. Water and air are the two substances in which this occurs. What actually happens is that localised heating, usually at the bottom, causes expansion and reduced density. The water or air so changed is raised by the entry of denser water or air into the lower position, and so a circulation occurs which moves heat rapidly through the mass. Air, when trapped, is a good insulator and this fact is used in such examples as hollow bricks for floor insulation and string vests for athletic 'types'. But air free to move will rise when heated so removing heat from the site of the heat source; a principle used in buildings relying on natural ventilation.

Effects of heat

When any material is heated, several changes obviously occur and some of these are significant in practical agriculture. They are: change of temperature, change of volume, and change of state; all are related to each other but will be dealt with separately.

Temperature change

This is the experience which is so familiar when any object is heated. One of the less obvious facts is that the temperatures of all substances do not change by the same amount for similar heat energy increases. If similar quantities of iron and water are heated the same amount, the temperature increase of the iron will be some nine times that of the water. The reaction of a substance to heat is calibrated as its specific heat, being compared to water having a specific heat of one. The specific heat of a substance is defined as the amount of heat required to raise the temperature of one gramme of the subtance one degree Celsius. Common examples are:

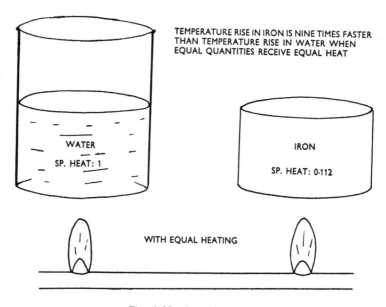

TEMPERATURE RISE IN IRON IS NINE TIMES FASTER THAN TEMPERATURE RISE IN WATER WHEN EQUAL QUANTITIES RECEIVE EQUAL HEAT

WATER
SP. HEAT: 1

IRON
SP. HEAT: 0.112

WITH EQUAL HEATING

Fig. 1.35   Specific heats.

Iron: 0.112   Copper: 0.095
Dry loam soil: 0.18   Wet loam soil: 0.53

It will be realised that the presence of water in a badly drained soil will raise its specific heat, with the effect of slow temperature rise in the spring. Temperature is measured by a thermometer which uses the expansion of a liquid as an indication of temperature.

(b) Thermometers

Various liquids may be used in thermometers, depending on the temperature range over which the instrument is designed to work. The two common liquids used are mercury, which appears silver, or alcohol which may be colourless but is often coloured red. The liquid is contained in a bulb at one end of an evacuated and sealed capillary tube. To calibrate the thermometer the bulb is immersed in water containing melting ice and the position at the top of the liquid in the capillary tube is marked on the stem. The thermometer is then placed so that the bulb is positioned just above the surface of boiling water and the level reached by the liquid in the thermometer is again marked. The lower mark is then designated 0° and the upper mark 100° when the Celsius scale is used. The stem is then divided along its length into 100 equal degrees. The old Fahrenheit scale, still found on many thermo-meters and instruments, designated 32°F as the melting point of ice and 212°F as the boiling point of water. There were 180 equal degrees between these two points. Interchange between the two scales is mathematically simple if it is remembered that 100°C equals 180°F, and that the Fahrenheit starts with freezing point as 32°.

Celsius Degrees = Fahrenheit Degrees $- 32 \times \frac{5}{9}$
Fahrenheit Degrees = Celsius Degrees $\times \frac{9}{5} + 32$

This general type of thermometer reads the temperature of the position wherever it may be sited, but in agricultural applications the extremes of temperature occurring over a period are often required. Firstly consider the need for a thermometer retaining the maximum temperature experienced. The common type of thermometer can, in manufacture, be modified to do this by producing a constriction in the capillary tube which allows the expanding liquid to pass but prevents its return on cooling. Such a thermometer is known as a maximum thermometer. The instrument can be re-set by shaking the liquid back past the constriction by centrifugal force.

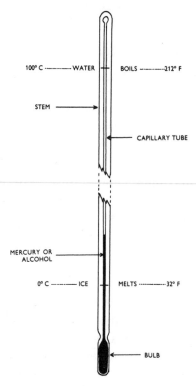

Fig. 1.36 Thermometer for general use.

Clinical thermometer

When a thermometer is used to find the temperature of an animal, the range of temperature is relatively narrow but considerable accuracy is needed; the clinical thermometer is, therefore, a special type of maximum thermometer. It is calibrated to read over a temperature range from 35 to 43°C on the Celsius scale or from 95 to 110°F on the old Fahrenheit scale.

When using a clinical thermometer, first ensure that the mercury has been shaken down and reads well below the anticipated body temperature of the animal. Next place the thermometer in or on the animal where it will achieve full body temperature, and leave it in position for the required time, usually a minute.

The stem of the thermometer is triangular in section, and to read should be rolled in the fingers to ensure that the thread of mercury is enlarged to an easily seen broad band.

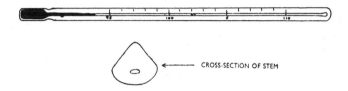

CROSS-SECTION OF STEM

Fig. 1.37 Clinical thermometer.

To record the maximum temperature experienced in a greenhouse or stockhouse, a similar type of maximum thermometer is used; the only difference being the wider range over which it is intended to work.

Minimum thermometers

These thermometers are alcohol filled and, in manufacture, a small dumb-bell shaped slider or index is inserted in the capillary tube. The thermometer is sited in a horizontal position and when this slider is immersed in the alcohol and the temperature is dropping, the surface tension on the alcohol surface in the capillary tube drags the slider back towards the bulb. As the temperature rises again, the alcohol flows past the slider, the upper end of which continues to indicate the lowest temperature reached until, by tilting the bulb upwards, the slider is again brought into contact with the alcohol surface.

Maximum and minimum thermometers

The fact that the foregoing types of maximum and minimum thermometers have to be removed from their fixing for resetting is a disadvantage in some applications; and in such situations as stockhouses a combined maximum and minimum thermometer, which can be permanently fixed, is used. This consists of a U-shaped thermometer in which the bulb and part of the stem is filled with alcohol, and the base of the 'U' is filled with mercury. In each limb of the thermometer, above the mercury level, is a small metallic slider fitted with a wire spring to grip the capillary tube causing the slider to 'stick' in position. The mercury merely acts as a spacer between the sliders on each side and, since the sliders float on the mercury, when the temperature drops the alcohol contracts towards the bulb; the mercury follows and raises the slider to the minimum temperature reached. When temperatures rise, the alcohol expands, flows past the slider and pushes the mercury around the bend of the 'U' which, in turn, raises the slider in the limb on the opposite side to the bulb to the maximum temperature reached. To reset, the sliders are pulled down in contact with the mercury by a small magnet.

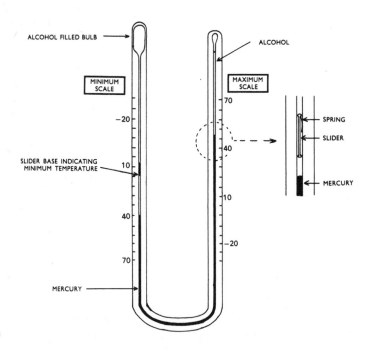

Fig. 1.38 Maximum and minimum thermometer.

(c) Change of state

If you are asked the question, what happens when an object is heated, the usual reply is that it gets hot; yet this is not always the case. Heating a substance does not always increase its temperature. Consider the effect of heating a container of ice blocks. If, at the start, the ice is colder than freezing point it will expand with heating and its temperature will rise. But as melting point is reached the temperature rise ceases and 0°C is maintained until all the ice is melted. The applied heat energy has been utilised in changing ice to water; this, in fact, amounts to using 334 joules to convert one gramme of ice to one gramme of water and is known as the latent (or hidden) heat of the fusion of ice. Supposing the container now containing water continues to be heated, the temperature rises steadily as time passes until boiling point is reached when a similar stand-still in temperature occurs. The temperature remains at boiling point until all the boiling water becomes steam, absorbing 2,260 joules to convert one gramme of water to steam—the latent heat of vaporisation, and if the container is enclosed then the temperature of the steam rises rapidly.

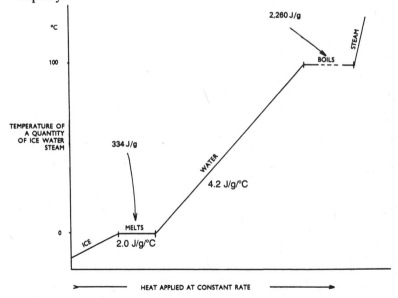

Fig. 1.39   Changes of state.

From the graph recording these temperature changes it will be seen that liquid water cannot exist below freezing point (under normal conditions) nor can it exist above boiling point. While water can only exist in the gas (or vapour) state above boiling point, the gas exists at all temperatures. Both ice and water evaporate to form water vapour in an attempt to saturate the immediately surrounding atmosphere. But this evaporation is still a change of state requiring the absorption of heat energy. This utilisation of heat energy when water evaporates is an important heat loss when a wet soil dries in the spring, resulting in slow temperature rise. It is also an important heat loss in stockhouses when floors or bedding are wet. We will consider this latter case in more detail later; but now study a situation where evaporation is required, namely in crop drying.

APPROXIMATE MOISTURE CONTENT OF GRAIN AT VARIOUS AIR TEMPERATURES AND RELATIVE HUMIDITIES

Air Temperature °C	Relative Humidity %								
	40	45	50	55	60	65	70	75	80
0	11.2	12.1	12.9	13.4	14.5	15.3	16.2	17.2	18.5
5	10.9	11.6	12.4	13.2	13.9	14.7	15.4	16.5	17.7
10	10.6	11.3	12.0	12.7	13.3	14.1	14.8	15.8	16.9
15	10.3	11.0	11.7	12.2	12.8	13.5	14.2	15.1	16.1
20	9.7	10.5	11.2	11.5	12.1	12.9	13.5	14.5	15.4
25	9.3	9.9	10.6	10.9	11.4	12.3	13.1	14.1	14.9

Safe storage conditions at figures below the line.

Grain drying

The moisture in the grain when harvested, at times amounting to 20 per cent by weight, in storage would promote seed activity, heating, and then deterioration by moulds and insects. For safe storage the moisture percentage must be reduced to 14 per cent; for every 1 per cent moisture reduction, 10 kg of water must be evaporated and removed from every tonne of grain. To evaporate 1 kg of water requires 2,550 kJ. Blowing air through the grain will produce a changing atmosphere which will eventually dry the grain and carry away the moisture but the process is very much more efficient if warm air is used. The latter acts as a heat carrier to the moisture in the grain, and has a lower relative humidity because it is warm; this also encourages the grain moisture to evaporate.

If the temperature of air is raised by 5°C, at 60 per cent relative humidity, it takes 750 m^3 of air to evaporate 1 kg of water.

Grain chilling

While one way of keeping grain is to dry it, there are some situations where, when it is used, the presence of moisture is an advantage. It may be useful to store grain wet, and deterioration by seed activity and undesirable organisms can be prevented by lowering the temperature.

This technique is facilitated by the fact that granular materials such as grain are self insulating and, providing that the internal atmosphere in a mass of grain is stationary, temperature changes of external conditions will influence only the outer few inches of the grain.

To prevent deterioration in grain above 22 per cent moisture requires temperatures approaching freezing point and is difficult and uneconomic: grain at 15 per cent moisture will keep for sixty weeks at 5°C, and thirty-two weeks at 15°C; while grain at 22 per cent will keep for six weeks at 5°C but for only one week at 15°C. The initial chilling which needs to be done immediately the grain is harvested, can be done by a relatively slow air movement (1–2 m/min) of refrigerated air through ducting on the floor under the grain which can be in a heap up to 6 m deep. To chill one tonne of 21 per cent moisture grain from 20°C to 5°C requires the removal of 30,000 kJ. In hot weather re-cooling is needed every ten to fourteen days, while through the winter re-cooling is only necessary every one to three months.

Refrigeration

In order to cool any object or any container, some means has to be found of moving heat from one place to another in opposition to the natural law that heat moves until all proximate objects are at the same temperature. In other words, a heat pump is needed. Examine the temperature graph of heated ice (fig. 1.39) again. From melting point onwards the water and then steam both expand considerably as heating progresses. These two reactions are rigidly linked together, so that as an object acquires more heat energy, it occupies a greater volume. An object or quantity of substance cannot change its heat status without an accompanying change in volume. You cannot prevent an iron bar expanding when it is heated, nor can air be compressed in a bicycle pump or compressor without producing heat.

When these changes occur in a substance they are most spectacular when change of physical state occurs and no change in temperature need occur. When 1 gramme of water at 100°C changes to steam it acquires 2,260 joules of heat energy and considerably increases in volume. Just as it is possible to control heat and produce volume change, so it is also possible to control volume and produce heat change.

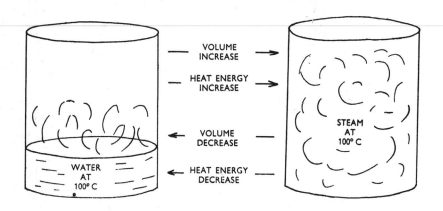

Fig. 1.40 Volume and heat energy changes with change of state.

If a chemical is chosen whose change of state occurs at normal temperatures, such as sulphur dioxide, application of pressure causes it to liquefy and release of pressure allows it to vaporise. But between these two states, there is a considerable difference in the amounts of heat energy contained. When the gas is liquefied it can contain very much less heat energy, and heat is therefore given up to its immediate surroundings; while when the liquid under pressure is allowed to expand through a jet, then to change to a gas it must acquire heat energy which is derived from the immediate surroundings. If these changes occur as the chemical, known as the refrigerant, passes around a continuous circuit, heat will be removed from the point where expansion occurs and transferred to the point where the gas is compressed. So we have a 'heat pump' which can be used not only to extract heat from a container and disperse it into a large atmosphere, as in the case of the domestic refrigerator, but also a device which can collect heat from a large low temperature source such as a river or the soil and transfer it to and concentrate it in a building. For example, The Festival Hall in London is heated by River Thames water heat.

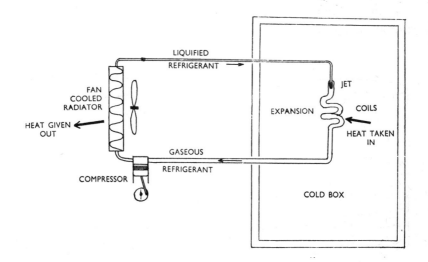

Fig. 1.41   Working principle of a refrigerator.

(d) Stock housing

Heat and stock housing

One of the basic facts about heat energy, mentioned in the section on heat movement, is that heat moves in various ways, from one place to another or from one object to another until the temperature of proximate objects is the same. A state of temperature equilibrium is naturally established.

Since farm animals are warm blooded—that is, the body temperature, in the region of 37°C, remains constant—in most situations the different temperature of the environment results in heat transfer, either to or from the animal. In addition to heat movement between the constant temperature animal and the variable temperature environment, the animal's metabolic processes result in some of the animal's food energy appearing as 'waste' heat. So the animal has its own internal heat source, amounting to 300–500 kJ/hr per 50 kg liveweight. The situation is summarised in the following diagram:

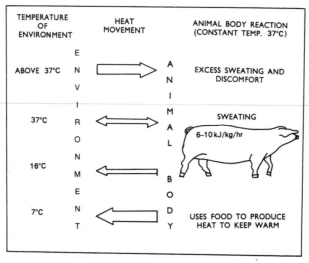

TEMPERATURE OF ENVIRONMENT	HEAT MOVEMENT	ANIMAL BODY REACTION (CONSTANT TEMP. 37°C)

ABOVE 37°C — ENVI — ANI — EXCESS SWEATING AND DISCOMFORT

37°C — RON — MAL — SWEATING — 6–10 kJ/kg/hr

16°C — NME — BO

7°C — ENT — DY — USES FOOD TO PRODUCE HEAT TO KEEP WARM

Fig. 1.42 Heat movement between an animal and its environment.

The temperature conditions under which stock are housed are important because, as any experienced stockman knows, uncomfortable animals are poor producers. The discomfort of animals can have a more direct economic effect. The purpose of feeding food to an animal is to produce meat or milk, and this purpose will be thwarted if the animal is cold because, if necessary, the animal's system will utilise some of its food to produce heat to maintain body temperature. The environmental temperature below which an animal 'burns' food to keep warm is known as the *lower critical temperature*. The class of stock and age influence this temperature as will be seen from the following table:

HOUSING OF FARM LIVESTOCK

Stock	Critical temp. range °C	Optimum temp. range °C	Heat pro- duced kJ hour	Water pro- duced litres hour	Space required m²	Ventilation rates m³/min	
						winter	summer
Milking cow	1–25	10–15	1900	0.3	2.75 shed 5.5 yard	1.4	5.6
Young calf	13–25	15	200		1.9	0.2	0.6
Piglets	15–27	27	30	0.0008			
Fatteners 50 kg	15–25	15–25	200	0.03	0.5	0.1	1.2
Bacon pigs	18–23	15–25	400	0.07	0.5		
Sows	10–25	10–15	800	0.1	2.75	0.4	1.7
Chickens Day-old		30–35	6				
Broilers 1 kg	13–30	15–21	20	0.006	0.06	0.008	0.06
Laying hens	7–31	10–13	40	0.01	0.1 slats 0.2 deep litter	0.06	0.3

Adult cattle, for instance will rarely use food for extra heat production, but fattening pigs will use 40 grammes food per day for every 1°C, their environmental temperature is below their lower critical temperature. In a 200-pig house this amounts to over 50 kg food per week, and the economic consideration is whether it is cheaper to feed extra food or insulate the house.

Similarly, if animals are too hot, in an attempt to reduce the heat produced by metabolism, intake and appetite will be reduced. Therefore production of meat or milk will be reduced. There is thus an *upper critical temperature* above which this occurs.

Heat losses

Since an animal produces its own heat, it is often possible to retain this heat in its immediate surroundings by design and insulation, and maintain comfortable conditions. Consider the ways in which this animal-heat travels to the outer atmosphere.

The animal body radiates heat to the surrounding solid structures which, in turn, pass the heat out to the atmosphere at a speed which depends on their construction. The air in contact with the animal becomes heated and rises by convection to carry heat

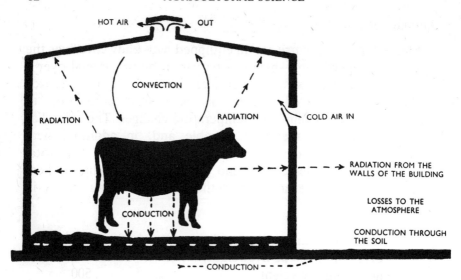

Fig. 1.43   Heat movement in stock housing.

out into the atmosphere through the ventilation system. When the animal lies down heat from its body travels, by conduction, through the floor and soil to the atmosphere.

A further important loss of heat in any stock house is the heat used up when moisture changes to water vapour when evaporation occurs from excreta and moist bedding. Using 2,260 kJ per kg of water evaporated, evaporation would theoretically cease when the internal atmosphere reached 100 per cent humidity; but under these conditions there would be no ventilation, and lack of fresh air and high humidity lead to respiratory problems. The design of the housing and its ventilation are obviously going to influence this method of heat loss.

The situation regarding the maintenance of comfortable stock housing conditions is therefore this: as much as possible of the heat produced by the animal must be retained within the building, by design and insulation, and the levels of internal temperature and humidity controlled by varying the speed of air movement through the house. This is oversimplifying the case but it will enable us to understand the interacting factors. Now consider these factors separately.

Animal body heat

Cattle, sheep, and poultry are all equipped with body coverings the insulating properties of which can be varied by the animal over a limited range. Pigs are poorly insulated creatures, but research workers have recorded changes in body heat production to compensate for environmental temperature changes. The animal as a heat producer is therefore variable and, in addition, heat production is not exactly proportional to body weight. The total heat output of various classes and ages of stock, including heat used to evaporate the moisture in their breath is given below in kJ/hr:

Cattle (according to liveweight)	2,500–3,500
Calves (up to 50 kg liveweight)	200– 350
Pigs: 50 kg liveweight	500
Pigs: 100 kg liveweight	800
Pigs: 150 kg liveweight	1,000
Pigs: medium weight sow	1,200
Pigs: 10 newborn piglets	300
Poultry: 100 day-olds	560
Poultry: 100 2 kg hens	4,000

Between $\frac{1}{3}$ and $\frac{2}{3}$ of this heat energy is available to heat the surroundings; the remainder is utilised in the evaporation of water.

Heat retention

There are various features about any substance which influence its behaviour as a heat barrier or insulator. If the atmosphere on each side of the barrier is at the same temperature, no heat moves; but when these temperatures differ, then heat tends to overcome the barrier by passing through it from the higher to the lower temperature atmosphere. This heat passage varies proportionally with the temperature difference on each side, and can be stated as a heat amount per °C difference. The heat passing through a housing structure also depends on the areas involved and is expressed as a quantity per m^2.

As far as the actual materials involved, the heat absorbing power of the internal surface, the heat conduction and thickness, and the radiation from the outer surface, all influence the insulating value. These variables are conveniently expressed as a single figure for commonly used forms of construction. This figure is the thermal conductance, or coefficient of heat transfer, being the number of joules which will pass through one square metre in one second for every degree Celsius temperature difference from one side to the other.

Most insulation materials make use of the fact that air is a very poor heat conductor. This property is only of value if the air is prevented from moving by convection currents. Any air spaces must therefore be small. Air can also carry water vapour, and heat conduction of any material is increased when it becomes damp or wet; so any air spaces in insulated construction must be sealed, and a vapour sealed surface is essential for any structure exposed to moist air to prevent condensed moisture dampening insulated walls and roofs.

Floors

Heat loss is by conduction into the floor fabric, laterally through the soil and thence to the atmosphere. The larger the building, the greater lateral distance the heat has to travel, therefore the thermal conductance decreases with increasing size. The natural drainage of the soil beneath the site also makes a considerable difference. For a building of which half the floor area is insulated by bedding or insulated construction, standing on a reasonably dry site, the following figures apply:

Building of floor area: 930 m^2 0.23 thermal conductance
233 m^2 0.4   „   „
58 m^2 0.6   „   „
9 m^2 1.1   „   „

Thermal conductance measured in $\text{J/m}^2\text{s}°\text{C}$. Low numbers show good insulation properties.

When constructing an insulated floor, a trapped air space of 20 mm should be provided, moisture sealed above and below. The flooring surface layer in contact with the animal should be kept as thin as possible to avoid delay in heating up when the animal changes position.

Walls and roofs

The insulating properties of these structures are influenced not only by the straightforward temperature difference between inside and outside the building, but additionally by the exposure of the site. A severely exposed building is unsheltered from the prevailing cold winds—usually from the north, north-east and east—while a sheltered site is one in which the force of the wind is broken by the lie of the land or by adjacent buildings or trees. Protection is afforded even when sheltering trees and buildings are not in front of the insulated house because exposure effects are due to general wind currents. The thermal conductance figures given below are for buildings of average exposure; for sheltered buildings they should be decreased by 12 per cent; and for buildings in exposed positions, should be increased by 12 per cent.

THERMAL CONDUCTANCE OF COMMON BUILDING CONSTRUCTIONS
low figures indicate good insulation
figures measured in $J/m^2s°C$

Brickwork 100 mm	3.4	Brickwork 200 mm	2.7
Cavity brick wall 250 mm	1.7	Cavity wall plus 25 mm cork board inside	0.9
Breeze block 200 mm	1.5	Concrete 100 mm	4.0
Corrugated asbestos	6.6	Asbestos plus 50 mm compressed straw	1.1
Flat asbestos	5.1	Corrugated iron	8.5
Single glazed window	5.7	Double glazed window	3.7
Wooden door 50 mm	2.3	Corrugated roof lights	6.8
Tiles	8.5	Tiles plus felt and 12 mm fibre board	1.7

From these figures it is possible to assess the relative insulating properties of different materials; and also the heat loss from a building may be calculated.

Ventilation

In considering ventilation, we continue the assumption that outside temperatures are low winter temperatures, and the interest is to maintain warm internal conditions. From what has been considered already, the heat loss through every m^2 of a building's construction, it might seem desirable to have as small a building as possible. Obviously there is a minimum floor space for any size animal, which dictates the minimum overall size; but since air movement is necessary to control humidity and temperature, an air flow of given rate which is imperceptible in a large volume building becomes a howling gale in a building of small total air volume. A howling gale is an exaggeration, but even draughts must be avoided for stock to be comfortable, and sensibly designed air inlets can avoid some of these problems.

Controlled air flow is the simplest way of maintaining a desirable internal temperature. Heating up air uses 32 kJ to heat 25 m^3 of air by 1°C; so when the outside temperature on a raw winter's day is 5°C, every 1 m^3 of air entering the house requires 13 kJ to warm it up to 15°C. As outside temperatures drop and the need to reduce heat loss becomes greater, it appears a simple matter to reduce the amount of air passing through a building. Unfortunately increasing humidity soon upsets this simple plan. The control of humidity is a much more important function of ventilation than temperature control.

Humidity

When humidity, that is the water vapour load in the air, is high—in the region of 80 per cent—then condensation is very inclined to occur on any cool surface, even if insulated. In addition, in high humidity with high temperatures, an animal's water vapour loss from the body is restricted and the animal finds difficulty in losing excess body heat; in high humidity and low temperatures, the air dampness increases the sense of cold and chilling is likely.

The only way humidity can be kept below 80 per cent is to allow sufficient air flow to remove the water vapour produced from the animal's breath and excreta.

If the internal temperature of a house is 15°C, and the external temperature is only 5°C, then it takes the following air amounts to carry away all the moisture produced by the stock per hour:

VOLUME OF AIR REQUIRED TO REMOVE MOISTURE VAPOUR FROM STOCK EACH HOUR

Outside air humidity per cent	70	80	90	100
Water vapour absorbed grammes/m^3	0.08	0.07	0.06	0.05
Large cow	45 m^3	49 m^3	56 m^3	68 m^3
Medium sow	15 m^3	16 m^3	18 m^3	23 m^3
Pig 100 kg	11 m^3	12 m^3	14 m^3	17 m^3
Chicks—100 day-olds	6 m^3	6 m^3	7 m^3	8 m^3
Hens—100	28 m^3	30 m^3	34 m^3	42 m^3

If in the house, drainage is poor and drinkers keep the floor wet then these figures may be increased; while good drainage and design would reduce the minimum air flow per hour.

Should the point be reached when the air flow necessary to maintain internal humidity below 80 per cent cools the house too much, then auxiliary heating of the incoming air becomes necessary. If the internal temperature in a house drops, even more air flow through the house becomes necessary to carry away the same amount of water vapour.

Methods of ventilation

Natural air movement in any location is caused either by winds or by convection currents. The cheapest form of ventilation makes use of these natural air movements to achieve air change in a building. Such ventilation will occur if openings are positioned on opposite sides of a house allowing wind to blow through; openings at the lowest and highest points on a house will produce the maximum convectional movement. The object is to control air flow, but these two forms of air movement are very difficult to control. From fig. 1.44, it will be seen that wind propelled air will produce draughts on the stock.

Design of variable openings and correct height-positioning can overcome some of this difficulty, but wind is a constantly changing force and it is impossible practically to achieve any reasonable control of the air-flow. Convection currents are equally as fickle; they are adequate so long as they are not upset by wind effects and

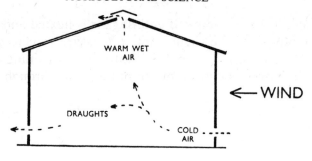

WARM WET
AIR

← WIND

DRAUGHTS

COLD
AIR

NATURAL VENTILATION BY CONVECTION

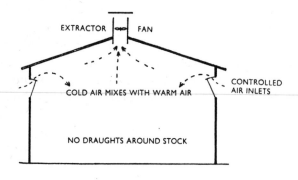

EXTRACTOR   FAN

COLD AIR MIXES WITH WARM AIR

CONTROLLED
AIR INLETS

NO DRAUGHTS AROUND STOCK

CONTROLLED VENTILATION

Fig. 1.44

conditions are not extreme. On a sharp frosty morning, when ventilation needs to be reduced, the internal and external temperature differences are greatest and so convection currents are really too effective. Conversely, on a foggy humid day, when maximum ventilation is needed, little or no convectional movement occurs.

The contrast to unreliable natural ventilation is forced ventilation by fans. Propeller fans are usually used and, since they are sensitive to back pressures produced by wind, the positions where these back pressures are least are the best working positions for fans. The ideal is at the roof ridge of a pitched roof building, or just above roof level on a flat roofed building. Ridge fans may be extractors assisting natural convectional ventilation; or they may be used as inlet impellers, this being most successful with artificially heated systems. If entering air is heated and forced into the building, then the slight positive pressure produced inside the building effectively prevents the entry of cold draughts and any other air movement is warm air passing out through cracks and small openings.

Summer ventilation

In summer the pressing need is to lower internal temperatures relative to outside temperatures and, since insulation is a barrier to heat movement irrespective of direction, a well insulated house will be cool in summer. It should be assumed that no heat will now be lost through the fabric of the building, and all heat has to be removed by air-flow. This means increasing the summer ventilation by four to six times. Water evaporation is now a useful way of reducing temperature and under extreme conditions some stock-men swill down the floors with water to cool the building.

Ventilation may therefore be summarised by stating that the minimum air-flow must be capable of dealing with the water vapour produced by the stock, and the maximum air-flow must restrict rising temperatures in summer.

2 Applied chemistry

(A) Elements, atoms and molecules

The earth is composed of about one hundred different pure basic building materials called elements. Each element has been given a name and a symbol.

The smallest part of an element which can partake in chemical activity is called an atom. The smallest particle which can exist alone is called a molecule. Separate molecules of hydrogen consist of two atoms, but there is only one atom in a carbon molecule.

Atoms of different elements can combine together to form compounds. A compound is different from either of its constituent elements, e.g. sodium chloride has different properties from sodium metal and chlorine gas from which it can be made. The smallest particle of a compound must contain at least two atoms, one from each constituent element, so it must be a molecule.

The proportions in which chemicals join together to form their compounds has taught us a lot about their combining power. Water is made from two parts of hydrogen gas and one part of oxygen gas by volume. Thus the combining power of oxygen must be twice that of hydrogen, and also they have opposite attractive forces since they join together.

The weight per volume, density, of an element informs the chemist of what size of atom the element consists.

From facts like these and the similarities between certain elements, chemists have built up a complex picture of the atomic strucure possessed by the elements.

The structure of an atom

Atoms have a dense core surrounded by orbiting rings of tiny particles called electrons.

The core is composed of positively charged protons, and neutrons without any charge. The electrons each carry an electric negative charge which is equal to the charge of one proton. The number of protons in the core is always equal to the number of electrons orbiting round it. Thus an atom has no external charge.

Hydrogen is the lightest known element with the smallest atom.

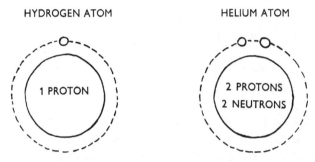

Fig. 2.1 Hydrogen and helium atoms.

The core of hydrogen has one proton in it and this is neutralised by one orbiting electron.

The next element is helium. This has two protons in its core, (and two neutrons), which are neutralised by two electrons in orbit. With two electrons orbiting, there seems to be no room left for a third so another orbiting sphere is begun further from the core.

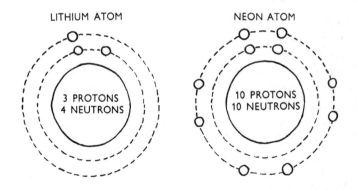

Fig. 2.2 Lithium and neon atoms.

This orbit is added to until there are eight electrons in it as in the neon atom. The orbit is then full and in this state seems to be stable, since neither neon nor helium will react chemically with other elements. The next orbit is begun with the sodium atom and is continued until there are again eight electrons present when the orbit is filled.

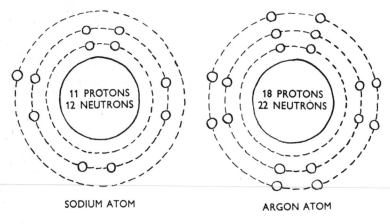

SODIUM ATOM   ARGON ATOM

Fig. 2.3 Sodium and argon atoms.

Electrons are interchangeable, negatively charged particles which can pass from one atom to another.

(B) Formation of ions

Ions are electrically charged chemical particles. When an atom of sodium meets an atom of chlorine there is an exchange of one electron.

The sodium atom reverts to the stable electron structure of neon by shedding its outer electron. But now there are eleven positive protons in the core and only ten negative electrons around it. The particle carries one extra positive charge and has become the sodium ion, Na^+. The chlorine atom gains an electron to assume the electron structure of argon which is stable. There are now eighteen negative electrons circling seventeen positive protons, so the particle carries one extra negative charge and is called the chloride ion, Cl^-.

If a solution contained sodium and chloride ions and a direct electric current were passed through it, the sodium ions, being positively charged, go to the negative pole or cathode and the chloride ions, being negative, go to the positive pole or anode.

A) FORMATION OF THE SODIUM ION

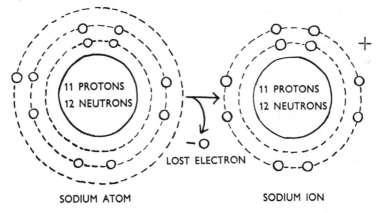

SODIUM ATOM   SODIUM ION

B) FORMATION OF THE CHLORIDE ION

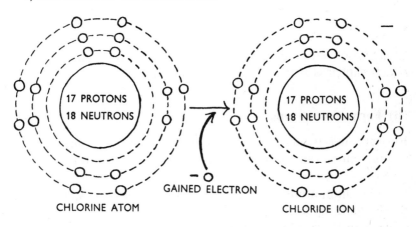

CHLORINE ATOM   CHLORIDE ION

Fig. 2.4

Positively-charged ions are therefore called cathode-seeking ions or cations as an easy abbreviation, and negatively charged ions are called anode-seeking ions, or anions.

Atoms of elements with two solitary electrons in their outside orbits can lose both of them to become double positive ions like calcium which forms the Ca^{++} cation, or magnesium which forms the Mg^{++} cation.

Similarly those elements whose atoms are short of two electrons to complete their outside orbit, can gain two more to become double negative anions. Oxygen forms the $O^=$ anion.

The number of charges which an ion carries determines its combining power. One sodium Na^+ combines with one chloride Cl^-, but one calcium Ca^{++} combines with two chloride Cl^-.

Sometimes complex ions are formed when groups of atoms carry an overall charge, e.g. the nitrate anion NO_3^- or the ammonium cation NH_4^+.

Some important soil cations:

Calcium	Ca^{++}	Sodium	Na^+
Magnesium	Mg^{++}	Ammonium	NH_4^+
Potassium	K^+	Hydrogen	H^+
Iron (Ferric)	Fe^{+++}	Aluminium	Al^{+++}

Important soil anions:

Nitrate	NO_3^-	Phosphate	$PO_4^\equiv$
Carbonate	CO_3^-		

(C) Solutions

True solutions

When a solid dissolves in water the resulting liquid is called a solution (aqueous solution). The solid is said to be water soluble. If the solid does not dissolve in water it is called insoluble.

The rate at which a solid dissolves in a liquid is generally increased as the solid is more finely ground or the temperature of the liquid is increased.

Fineness of grinding is important when plant fertilisers are only slowly soluble in water. If basic slag, calcium carbonate (limestone) and bone meal are not finely ground before application to the soil, their rate of action is too slow.

Solute and solvent

In a true solution the liquid is transparent, although it may be coloured. The dissolving liquid is called the solvent, and the substance dissolved is the solute.

Solvents can be liquid or gas; and solutes can be solid, liquid, or gas.

Saturation

When no more solute will dissolve in a solvent at a certain temperature, the solution is saturated. Cooling a saturated solution renders it super-saturated and crystals of pure solute will form. A good example of this happening is air saturated with water vapour. The water is invisible in air until the air is cooled, it then becomes super-saturated with water and water droplets appear in the air causing fog. Often saturated air is cooled on the ground or on cold walls causing dew or condensation to form. Therefore, keeping the air warm prevents condensation in a building.

Miscible liquids

If two liquids are mutually insoluble, they are said to be immiscible, e.g. water and petrol. Even if they are shaken together they soon separate to form droplets which gather together into separate liquids again.

Paraffin and petrol are so similar in structure that they mix easily in all proportions.

Emulsions

If soap is added to an oil and water mixture, it acts as a stabiliser and prevents the droplets of oil coming together again. A stable suspension of oil drops in water is called an emulsion.

Emulsions are very important in agriculture as many of the selective weedkillers used are oily solutions which must be diluted in water before use. The manufacturers emulsify the spray before it is marketed.

The soap solution in the spray helps the water drops to spread more evenly over plant leaves by lowering the surface tension of the drops. Without these soaps or wetting agents, water drops stay in small beads on plant leaves.

Colloidal solutions

True solutions contain particles of solute of molecular size. Suspensions contain large particles of solute which usually settle out. Colloidal solutions contain particles of solute, larger than molecules but smaller than suspended particles.

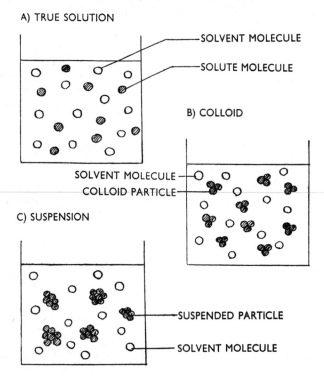

Fig. 2.5   True solution, colloid and suspension.

Stability of colloidal solutions

At the surface of the particles, positive or negative ions are adsorbed and held by strong forces. The particles are all coated with the same type of charged ion, i.e. all negative or all positive. The particles then repel one another as do similar poles on magnets.

Coagulation of particles

Adding a strong acid or alkali to the solution can displace the adsorbed ions so that the particles no longer repel one another and flocculate.

Passing a high voltage through the solution, or raising its temperature, also causes flocculation of the particles.

Clay suspensions

These often contain particles of colloidal size which react similarly to colloids:
1. Clay particles repel one another in solution.
2. Clay particles adsorb ions on to their surface.
3. They are suspended indefinitely (deflocculated) by monovalent cations like sodium.
4. They are coagulated (flocculated) by electrolytes.
5. They are coagulated by divalent ions like calcium and magnesium.

(D) Base exchange

Base exchange or cation exchange in clay soil

Sand particles carry no electrical charges, so they are useless for holding mineral ionic particles in the soil against gravity.

Clay and humus particles are negatively charged. In the soil they become surrounded by cations which are dissolved in the soil water. These cations can be exchanged for any others if the newcomers arrive in sufficient concentration. The ions displaced pass through the soil in the drainage water, i.e. they are leached.

As seen in the fig. 2.6, divalent cations like calcium, having two positive charges, can form links between particles and bring about flocculation.

Negatively charged anions cannot be held by the soil, so nitrate (NO_3^-) should always be applied to the growing crop to avoid leaching losses. Applications of nitrate fertiliser to the seed bed of a winter cereal can be wasteful, as the plants only use a little before winter and any surplus is leached before the spring.

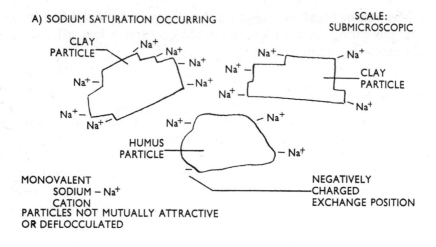

A) SODIUM SATURATION OCCURRING

SCALE: SUBMICROSCOPIC

CLAY PARTICLE

CLAY PARTICLE

HUMUS PARTICLE

MONOVALENT SODIUM – Na⁺ CATION

NEGATIVELY CHARGED EXCHANGE POSITION

PARTICLES NOT MUTUALLY ATTRACTIVE OR DEFLOCCULATED

B) CALCIUM SATURATION OCCURRING AFTER LIME APPLICATION

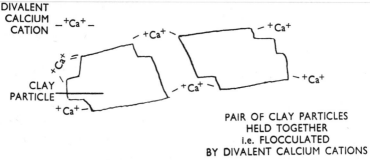

DIVALENT CALCIUM CATION

CLAY PARTICLE

PAIR OF CLAY PARTICLES HELD TOGETHER i.e. FLOCCULATED BY DIVALENT CALCIUM CATIONS

Fig. 2.6   Base exchange.

Ammonium sulphate and soil acidity

Ammonium sulphate and ammonium nitrate fertilisers can be responsible for increasing soil acidity. Ammonium sulphate and ammonium nitrate are acid salts, i.e., a combination of a weak base and a strong acid. The removal of the ammonium ions from the soil leaves the strongly acid sulphate or nitrate ions to increase acidity. Ammonium cations can displace calcium from clay particles. The ammonium ions are attacked by nitrifying bacteria in the soil and oxidised to nitrate anions. Since the charge on the ion is changed by the chemical re-shuffle, the particle is now repelled by the clay.

The vacant negative charge on the clay particle is satisfied by a hyrogen cation. A hydrogen-saturated clay has a low pH and is unsuitable for plant growth.

(E) Carbon dioxide

Carbon dioxide is the starting point in the chain of organic compounds which build plants and animals.

The endothermic chemical action occurring in green plants which changes carbon dioxide and water into carbohydrates is called photosynthesis. Photosynthesis can only occur in those parts of living plants which contain the green pigment chlorophyll, and which receive adequate sunlight.

Equation for photosynthesis:

$$nCO_2 + nH_2O \longrightarrow (CH_2O)n + nO_2$$

carbon   water   carbo-   oxygen
dioxide   hydrate

(n is an unknown number but probably 3)

Although carbon dioxide is very important, it is only found as a gas in air at 0.03 per cent, or 3 parts in 10,000. Because of its scarcity, the potential rate of photosynthesis is restricted in most years by this factor.

Raising the level of carbon dioxide concentration to about 1 per cent of the air in glasshouses has brought about substantial increases in crop growth. Above 1 per cent, the carbon dioxide depresses the growth rate of many plants.

Properties of carbon dioxide (CO_2)

It is a colourless, odourless, acidic gas which is heavier than air. It is soluble in water to produce a weak acid, carbonic acid (H_2CO_3). Carbon dioxide does not support combustion or the respiration of living organisms. For this reason it is produced in a fire extinguisher by the action of hydrochloric acid or sulphuric acid on calcium carbonate, to quench flames. A good test for carbon dioxide is its ability to put out a flame or to turn lime water ($Ca(OH)_2$) milky.

Respiration and carbon dioxide production

The respiration of animals or stored plant products like grain or potatoes, produces carbon dioxide gas. The bacteria and fungi in the soil produce most carbon dioxide, maintaining the atmospheric supply.

Equation for respiration:

$$(CH_2O)n \; + \; nO_2 \longrightarrow nCO_2 \; + nH_2O$$

carbohydrate   oxygen   carbon   water
                                       dioxide

It could be dangerous for a man to climb into a silo on top of grain where invisible carbon dioxide might have collected. Carrying a match or candle alight is a useful precaution. If the candle goes out, get out as quickly as possible.

Animal houses which are badly ventilated can cause carbon dioxide to build up to unhealthy levels. High levels of carbon dioxide increase the breathing rate of animals.

As carbon dioxide forms an acid with water, it is partly responsible for atmospheric corrosion of metal.

(F) Oxygen and oxidation

Oxygen

Oxygen is a gas found abundantly in fresh air at 20 per cent by volume, or 1 part in 5. This level is maintained by constant production of oxygen gas by green plants in photosynthesis. This production of oxygen equals its rate of absorption by various chemical oxidation reactions.

Properties of oxygen

Oxygen is a colourless, odourless gas, supporting combustion and respiration.

Test for oxygen

A glowing splint is re-ignited by pure oxygen gas.

Oxygen and respiration

Respiration is an exothermic chemical action occurring in all living organisms. Most organisms require oxygen to respire, although a few bacteria and fungi and some deep-seated body muscle cells can respire for a short time without oxygen. Anaerobic respiration, as this is called, produces alcohol or organic acids like lactic acid. These substances quickly inhibit growth when their concentration rises; a fact made use of in silage fermentation.

Equation for Anaerobic Respiration:

$$3(CH_2O)n \longrightarrow nCO_2 + nC_2H_5OH$$

carbohydrate   carbon   ethyl
                                  dioxide   alcohol

In animals the oxygen combines with the haemoglobin of the red blood corpuscles as they pass through the pulmonary arteries and blood capillaries. The oxidised haemoglobin called oxyhaemoglobin is bright red in colour are very unstable chemically. When this compound reaches the muscle cells, skin, or internal organs where respiration is taking place, the oxygen leaves the haemoglobin to take part in respiration. It should be noted that respiration is a cellular chemical action producing energy.

An adequate supply of oxygen is essential to maintain health, growth, and production in farm animals.

Oxidation

The chemical combination of oxygen with elements or the removal of hydrogen from elements.

Example:

The oxidation of ammonia to nitrate as found in the action of the soil nitrifying bacteria.

$$NH_3 \quad \overset{\displaystyle 3H}{\underset{\displaystyle 3O}{\rule{3cm}{0.4pt}}} \quad NO_3$$

ammonia   nitrate

This oxidation involves both the removal of hydrogen and the addition of oxygen.

Oxides

Oxygen combines with metallic elements like magnesium or calcium to form basic oxides which dissolve in water to form alkalis.

Oxygen combines with non-metallic elements like carbon, phosphorus, and sulphur to produce acidic oxides which dissolve in water to form acids.

(G) Acids, alkalis and pH

The oxides of non-metallic elements dissolved in water form acids e.g.:

(i)
$$C + O_2 \longrightarrow CO_2$$
$$CO_2 + H_2O \longrightarrow H_2CO_3$$
carbonic acid

(ii)
$$2P + 5O \longrightarrow P_2O_5$$
phosphorus   oxygen   phosphorus pentoxide
$$P_2O_5 + 3H_2O \longrightarrow 2H_3PO_4$$
phosphoric acid

Alkalis are the oxides of metals dissolved in water, e.g.

$$Ca + O \longrightarrow CaO$$
calcium   oxygen   calcium oxide
$$CaO + H_2O \longrightarrow Ca(OH)_2$$
calcium hydroxide

Equal amounts of similar strength acid and alkali will exactly neutralise one another, forming a neutral salt and water, e.g.

$$Ca(OH)_2 + H_2CO_3 \rightarrow CaCO_3 + 2H_2O$$
calcium hydroxide   carbonic acid   calcium
carbonate

Calcium hydroxide is very valuable for neutralising soil acidity.

pH

This is a numbered scale to measure acidity or alkalinity.

1 2 3 4 5 6 7 8 9 10 11 12 13 14

acid   neutral   alkaline

pH 1 is most acid, while pH 14 is most alkaline.

The important thing about this scale is that pH 6 is 10 times more acid than pH 7, and pH 5 is 100 times more acid than pH 7, and so on.

Plants can tolerate only slight pH variation from neutral.

(H) Metal corrosion

This is a process of electron transfer similar to that occurring in a primary electric cell. When two metals are in contact in the presence of impure water and oxygen, electrons are passed from the more active metal to the less active metal. The active metal will slowly corrode away: the inactive metal does not corrode.

Iron is a fairly active metal and will corrode rapidly in the presence of an acid which takes the place of the inactive metal.

Preventing corrosion

1. Keep the metal dry; the electrons cannot move without liquid so no corrosion takes place. This can best be achieved by greasing or painting the metal.
2. Attach to the metal another metal which is more active, e.g. magnesium or zinc is more active than iron. The active metal will deteriorate leaving the iron protected. This is the principle behind zinc galvanising of iron. Blocks of active metal can be strapped to a car chassis in strategic places to prevent corrosion.

(I) Water and water supplies

Most of the water collected for town water supplies has dissolved minerals on its way to the reservoir. The purification of water for drinking does not alter its mineral content.

Purification of water

While water is being held in a reservoir, the clay particles suspended in it are sedimented on the reservoir floor. Any organic matter floats to the surface and can be more easily filtered from the stream entering the water works.

The water is filtered several times through gravel to remove all traces of floating matter. To kill bacteria or other organisms in the water, chlorine is added at the rate of one part per million. If the water is slightly acid after treatment, lime is added because acid water can dissolve lead from water pipes.

Hard water

Calcium and magnesium bicarbonate salts present in water, are precipitated as insoluble calcium and magnesium carbonates on heating.

Hard water is unsuitable for making steam as the precipitates or scale settle on the boiler walls, and in the hot pipes. As the scale thickens, heat transfer is less efficient and eventually pipes can be blocked.

Water softening

Sodium carbonate (washing soda) precipitates calcium salts from water which is to be used for washing. Soap can then be used to greater effect.

Soap will soften water but it forms an insoluble scum which floats on the surface.

Detergents soften water and have the advantage that they form soluble compounds with the calcium salts in the water.

Zeolite water softeners

Zeolites are clay minerals. If hard water is filtered through a column of sodium-saturated clay particles, the calcium in the water changes place with the sodium cations attached to the clay. The water escaping from the clay has lost its calcium and gained soluble sodium salts in its place. When the clay column becomes saturated with calcium eventually, a concentrated brine solution is poured through to bring the clay back to sodium saturation.

Water softeners like this, using the base exchange principle, provide a useful supply of softened water for washing.

Water pollution

Chemical pollution of the environment in which we work, take our ease, and upon which we are dependant for our food, is one of the most serious problems of our industrialised society. Most people are aware of the dangers associated with the disposal of atomic waste or of the danger connected with the use of persistent organo-chlorine insecticides, but many are unaware of the serious consequences of disturbing the balance which exists in most natural environments.

The practice of applying soluble nitrate fertilisers to cereal crops and grassland in an attempt to maximise leaf development leads to leaching of the nitrates into drainage water in wet weather, and consequently to an increase of nitrate in stream and river water. The quantity of dissolved nitrate in some drinking water supplies is giving cause for concern. The levels are monitored regularly by the water authorities in case there should be a sudden increase.

The blocking of water courses by excessive water weed growth, as a result of nitrate pollution of water, is leading to a process called 'eutrophication'. First, reeds and bullrush grow on the banks of the stream or lake, then flotsam and silt build up, dying vegetation rots in the water, the small supply of dissolved oxygen is used up by the rotting process; and finally, without oxygen, all aerobic life in the river dies off.

The accidental pollution of farm water courses by slurry is a constant worry for intensive livestock farmers. The slurry encourages the growth of bacteria in the water, which causes an increased biological oxygen demand (B.O.D.) and it is not long before fish are unable to find enough dissolved oxygen, and die. Effluent from silage pits is about the worst offender as it is notorious for increasing the B.O.D. and can rapidly lead to a lifeless stream.

If we could replace excess nitrate fertiliser dressings by regular applications of slurry to cereals and grass, we could overcome two problems at the same time.

3 Cytology and inheritance

(A) Cell structure

All living plants and animals are made of a collection of minute cells which are only visible under a miscroscope. The structure, function and reproduction of these cells vary greatly. Their study is termed cytology.

Some cells retain a simple structure throughout their lives and probably perform diverse duties for the animal or plant of which they are part. Other cells become highly specialised and perform one complex action only for the organism. Complexity of structure and function usually develops as the cell grows. Young cells usually conform to a general pattern.

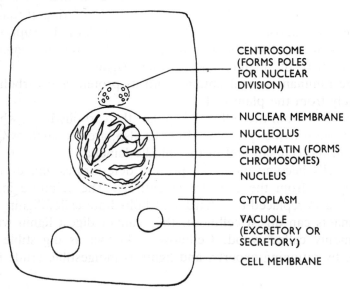

Fig. 3.1 General structure of cell.

The cell is enclosed in an outer membrane or wall which is made either of elastic protein, or of more solid carbohydrate cellulose.

Inside the membrane is a layer of protein cytoplasm or protoplasm, jelly-like in consistency and able to transport the contents around the cell by flowing. Within the cytoplasm are numerous dense inclusions, the most important of which is the nucleus. This dense protein mass controls the activity and inheritance of cell contents. Other inclusions are found: many have special duties to perform, like the chlorophyll-containing plastids of plant cells which trap sunlight for photosynthesis. Within the cytoplasm there may be a liquid-filled cavity called a vacuole. In plants there usually is a vacuole, but more often in animal cells the vacuole is absent.

(B) Plant cells

The outside layer of a plant cell is made of a carbohydrate called cellulose. This is precipitated onto a fine membrane of calcium pectate called the middle lamella. The cellulose wall is fully permeable to water but the cytoplasm layer within is only semi-permeable. The cell sap inside the vacuole of the cell is usually sugary, so the cell is blown up by water entering through the process of osmosis. Since herbaceous plants lack the support of a rigid skeleton, the inflation of cells to the turgid condition is important to the maintenance of plant rigidity.

The ruminant animal obtains both its protein and carbohydrate entirely from the plant cell.

Cellulose from the cell wall supplies the carbohydrate, while the cytoplasm provides protein. Minerals also, are provided by the cell both in the cell sap and in the middle lamella.

As cells become older the protein deteriorates and eventually disappears from the cell almost entirely. The cellulose in many cells is also replaced by a harder carbohydrate called lignin. While ruminants can digest cellulose, they cannot digest lignin which is commonly called wood. Cellulose is known as digestible crude fibre, by rationing experts, and lignin is indigestible crude fibre.

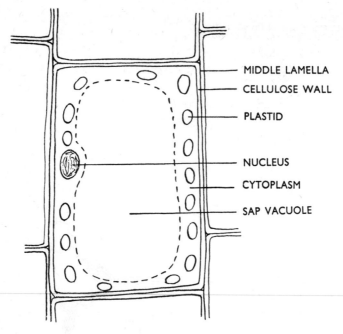

Fig. 3.2   Typical plant cell, photosynthetic leaf cell.

(C) Animal cells

The outer membrane is often difficult to see under a microscope, as it is thin and probably made of elastic protein. The cell is usually filled by cytoplasm, although there may be a vacuole appearing from time to time if a cell is excreting waste or secreting hormone, enzyme, etc. The nucleus is large and easily stained with basic dyes like haematoxylin which colours it purple.

Simple animals made of single cells or a few cells, have many non-specialised cells within their bodies which retain the ability to take in food, excrete waste, be sensitive, and reproduce. But as animals become more complex and contain more and more cells, these cells lose the ability to perform basic tasks like excretion, sensitivity, and reproduction. Special units of cells gather together to perform a function for the body, e.g. excretory cells are massed into a unit called the kidney, and sensory cells are massed into units called eyes and ears. The ability to reproduce is taken over by cells in the gonads (testes and ovaries), but there remain many

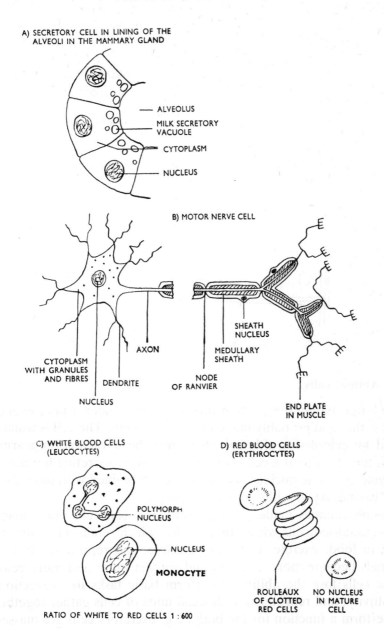

A) SECRETORY CELL IN LINING OF THE
ALVEOLI IN THE MAMMARY GLAND

— ALVEOLUS

MILK SECRETORY
VACUOLE

CYTOPLASM

NUCLEUS

B) MOTOR NERVE CELL

SHEATH
NUCLEUS

AXON

MEDULLARY
SHEATH

CYTOPLASM
WITH GRANULES
AND FIBRES

DENDRITE

NODE
OF RANVIER

NUCLEUS

END PLATE
IN MUSCLE

C) WHITE BLOOD CELLS
(LEUCOCYTES)

D) RED BLOOD CELLS
(ERYTHROCYTES)

POLYMORPH
NUCLEUS

NUCLEUS

MONOCYTE

ROULEÁUX
OF CLOTTED
RED CELLS

NO NUCLEUS
IN MATURE
CELL

RATIO OF WHITE TO RED CELLS 1 : 600

Fig. 3.3 Specialised animal cells.

centres in the rest of the body where cells divide for growth and replacement of worn out tissue. For example, the bone marrow is active in producing replacement red blood cells which deteriorate after about three months' steady use.

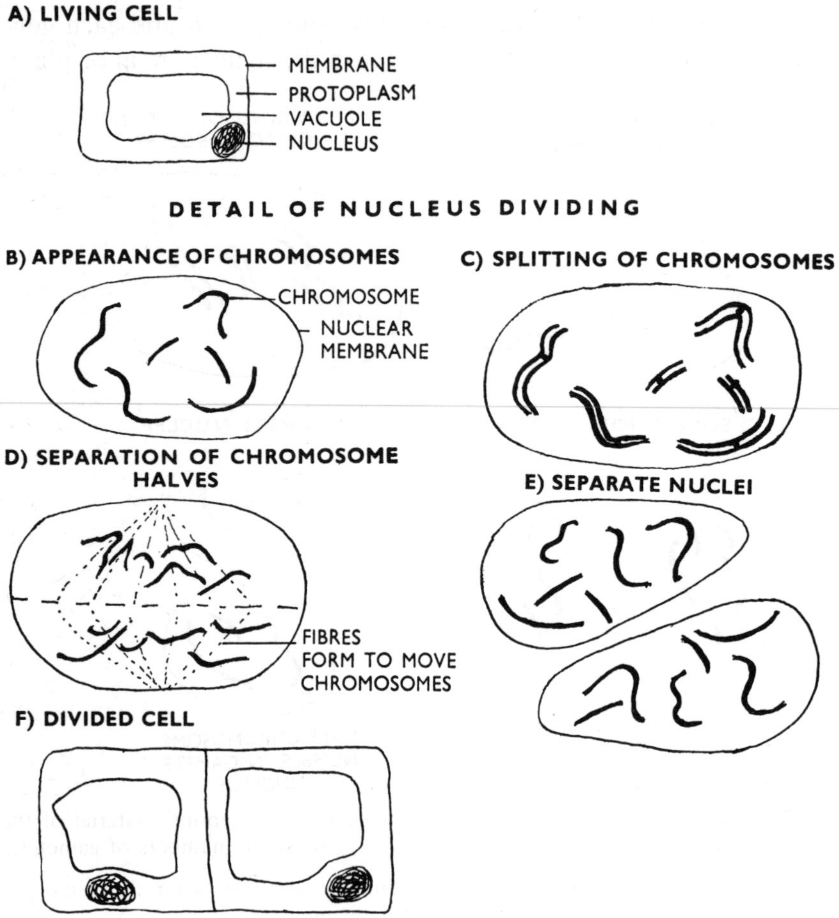

A) LIVING CELL

— MEMBRANE
— PROTOPLASM
— VACUOLE
— NUCLEUS

DETAIL OF NUCLEUS DIVIDING

B) APPEARANCE OF CHROMOSOMES

—CHROMOSOME
— NUCLEAR MEMBRANE

C) SPLITTING OF CHROMOSOMES

D) SEPARATION OF CHROMOSOME HALVES

FIBRES FORM TO MOVE CHROMOSOMES

E) SEPARATE NUCLEI

F) DIVIDED CELL

Fig. 3.4 Growth division or mitosis.

(D) Cell reproduction

Apart from controlling the growth and activity of the cell, the nucleus contains the 'blueprint' or design for the cell and its function.

This 'blueprint' is carried on the chromosomes of the nucleus. The chromosomes are long threads of protein called chromatin woven into a dense mass inside the nuclear membrane.

Chromosomes have been examined closely by magnifying them many thousands of times under electron microscopes. The chromosome band is made of sections called genes, and each gene has a controlling effect on a character in the living animal or plant.

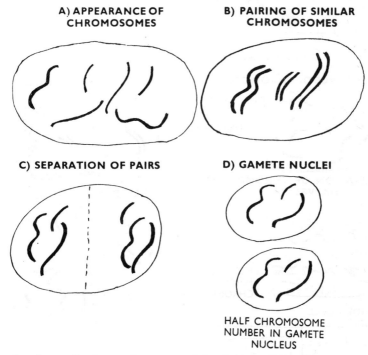

A) APPEARANCE OF CHROMOSOMES

B) PAIRING OF SIMILAR CHROMOSOMES

C) SEPARATION OF PAIRS

D) GAMETE NUCLEI

HALF CHROMOSOME NUMBER IN GAMETE NUCLEUS

The daughter cells shown above, containing half the 'blueprint' material of the parent, subsequently divide again exactly to increase the numbers of gametes.

Fig. 3.5 Reduction division or meiosis and formation of sex cells (gametes), detail of nucleus dividing.

Chromosomes are present in duplicate sets in all body cells of animals and most plants. Not only the chromosomes are duplicated, but also the genes. Therefore every nucleus of most animals and plants contains two genes which have a controlling effect on the same character.

When gametes are formed, i.e. sperm or ova in animals, only

one gene is present for each character so when they fuse to form a zygote, a duplicate set of chromosomes and genes is present in the nucleus. One gene has arrived in the new offspring from each parent.

Two types of cell division therefore must occur in living organisms.

One type produces exact copies of the cell, its nucleus and 'blueprint' contents. This is called mitosis and is part of the normal growth process.

The other type must produce special cells for reproduction, the gametes, which have half the 'blueprint' material of an organism. This division is called meiosis.

(E) Genetics and the work of Mendel

These relatively recent chromosome discoveries have thrown light on the mechanism of inheritance discovered and described by Gregor Mendel, an Augustinian monk.

Mendel carried out elaborate plant breeding experiments at his monastery in Brunn, Czechoslovakia, in the middle of the nineteenth century. His results were submitted to the leading biological societies of his day, but no importance was attached to them until the beginning of the twentieth century.

Mendel worked with the garden pea plant. First he selected a plant which showed a character clearly and bred true to type.

One such character was seed-pod colour. Here he found two clearly different pure breeding types. Those with yellow pods, and those with green. These pure breeding types he crossed by sexual reproduction, and carefully observed the pod colour in the first generation of offspring.

The important facts which Mendel discovered were:

1. The inheritance of characters is dependent on factors which were passed on from parent to offspring.
   We now call these factors genes.
2. The genes controlling a character are present in pairs.
   We now know that the genes are on pairs of chromosomes in the nucleus.
3. Only one gene of a pair can enter a gamete. This is

explained perfectly by our recent knowledge of chromosome behaviour during meiosis.

4. In any pair of genes affecting one character:
Either: The genes may be alike, having the same effect on the character. The organism is then pure breeding and said to be homozygous for that character.
A most important result of this is that the internal genetical structure or genotype of the organism is bound to be the same as the external appearance or phenotype of the organism.
Or: The genes may be different. The organism is then not pure breeding and is said to be heterozygous for that character.
The internal genetical structure or genotype of the organism can be different from the external appearance or phenotype of the organism.
This is especially true if the effect of the genes in a pair is not equal. If they are unequal the gene with the stronger effect is said to be dominant and the weaker gene recessive. The external appearance of the organism is determined by the dominant gene while the recessive is completely hidden. Sometimes different genes have equal effects. For example, a red flower crossed with a white flower can give a pink flower in the heterozygous condition.

We can easily underestimate the ability and knowledge of this monk, Gregor Mendel, unless we try as he did to carry out these crosses in an exact way.

He knew that pea flowers were normally self-pollinating. He had to prevent them from doing this, so he opened the flower buds and carefully removed all the stamens before the stigma was ready to receive pollen. These emasculated flowers were enclosed in muslin bags to prevent insects from interfering with Mendel's work. When the stigma of the flower was ripe, he collected pollen from the stamens on a plant of the opposite type and dusted it onto the ripe stigma.

Since he worked with about seven different characters on the pea plant, he must have had many separate plots of carefully named and recorded peas. Some would be parent lines, some first

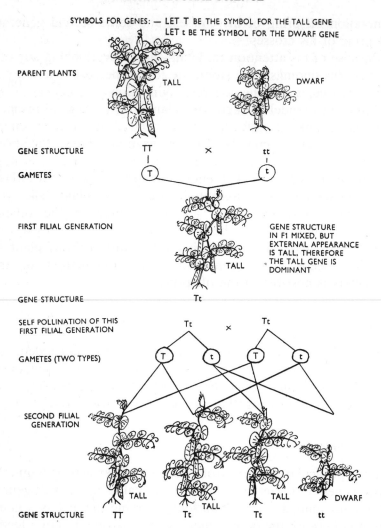

SYMBOLS FOR GENES: — LET T BE THE SYMBOL FOR THE TALL GENE

LET t BE THE SYMBOL FOR THE DWARF GENE

PARENT PLANTS   TALL   DWARF

GENE STRUCTURE   TT   ×   tt

GAMETES   T   t

FIRST FILIAL GENERATION   TALL

GENE STRUCTURE IN F1 MIXED, BUT EXTERNAL APPEARANCE IS TALL, THEREFORE THE TALL GENE IS DOMINANT

GENE STRUCTURE   Tt

SELF POLLINATION OF THIS FIRST FILIAL GENERATION   Tt   ×   Tt

GAMETES (TWO TYPES)   T   t   T   t

SECOND FILIAL GENERATION   TALL   TALL   TALL   DWARF

GENE STRUCTURE   TT   Tt   Tt   tt

Results. In external appearance there is a 3:1 ratio in favour of the dominant gene.

In genetical structure there are present
- *1 pure breeding tall*
- *2 hybrid talls*
- *1 pure breeding dwarf*

Fig. 3.6   Mendelian inheritance illustrated by a cross between a pure breeding tall pea plant and a pure breeding dwarf.

generation offspring of his crosses, and some second generation offspring of his crosses.

Because of his attention to detail—never overlooking any of his results—his findings have stood the test of persistent scrutiny and are now accepted as the concrete foundation of genetical science.

Although Mendel worked with plants, his findings in respect of dominance and recessiveness have been demonstrated since in many characters in animals also. One example here, is the absence of horns in cattle (polled) being dominant over presence of horns.

One final theory which Mendel put forward was that although many characters are present in the same plant, the genes controlling these characters are recombined in the offspring independently of one another. This means that if a tall plant with round seeds and yellow pods, is crossed with a dwarf plant with wrinkled seeds and green pods, any combination of these characters is possible in the offspring.

Linkage

Since Mendel's time, further work has shown that this independent segregation of genes at gamete formation does not always take place. Those genes which are found on the same chromosome often remain linked together during gamete formation.

(F) Sex linkage

On one particular pair of chromosomes, the sex gene is found. The other genes on these chromosomes are called sex linked genes.

In many animals the chromosome carrying the male sex gene is different because the other genes which it carries are not effective.

Whenever the male sex gene appears in an embryo, that embryo is a potential male. Only potential since hormones are essential for the animal's full development. Since only the male parent produces the male sex gene, the male parent determines the sex of the offspring. Also, since the male sex chromosome carries inactive genes, all the characters affected by the sex linked genes are controlled by the chromosome from the maternal parent.

Apart from an interesting observation that boys will always inherit certain characters from their mothers, there is one other disturbing effect. Any dangerous recessive genes, which would normally be hidden in an animal by a dominant partner gene, will control the character of the male offspring. This can account for a higher mortality rate in male offspring among mammals.

Birds, including poultry, differ from mammals in that the female sex chromosome is deficient in active genes. Therefore pullet chicks (females) look like the paternal parent in all its sex linked characters. This fact can be used to sex chicks at hatching. An example of how this can be achieved is as follows:

If a Rhode Island Red (pure breeding red feathered) hen is mated with a Light Sussex cockerel (pure breeding white feathered), all the chicks are white feathered when they hatch.

White feather colour is, therefore, dominant over red feather colour. However, the gene controlling feather colour is sex linked, so a white feathered hen carries only one white feathered gene. The other gene, being linked on the deficient female sex chromosome, carries no colour. If a Rhode Island Red cockerel is mated with a Light Sussex hen all the pullets will carry the cockerel's feather colour and will appear red. The cockerel chicks will all be white feathered like the hen parent.

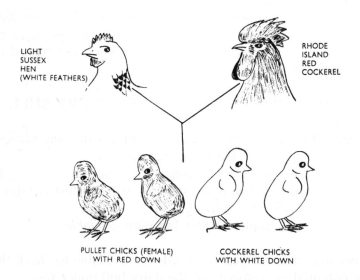

Fig. 3.7 Sex linked colour inheritance in poultry.

(G) Animal breeding

Inheritance can be a complicated process, as when characters like milk production are inherited, since this is affected by many genes at once. But an understanding of Mendelian principles can explain a few things in animal and plant breeding.

1. Pure breeding, homozygous lines are preferable when starting a breeding programme. To obtain these, a series of close matings must be carried out within families of animals. When breeding from crossbreds or hybrids with heterozygous genes, the results cannot be predicted.

2. The external appearance or phenotype of an animal can be misleading since it is not bound to be the same as its genotype or breeding capability.

If an animal is going to be used extensively in breeding, as would be a bull used by the Milk Marketing Board at one of their artificial insemination centres, then test matings should be carried out on a small scale first to check the genotype of the animal. This means that a dairy bull may be $5\frac{1}{2}$ years old before it has been initially proved (see table below).

The ability to collect semen from a bull and store it in deep freeze cabinets has proved valuable since a bull may prove to be extremely useful in a breeding programme when it has passed its best, or may even be dead!

TIME TAKEN TO TEST MATE A NEW DAIRY BULL

Age of bull

1.5 years	First test matings (can be earlier with some breeds)
2.25 years	Calves born from test matings
3.75 years	Heifer calves old enough to be mated
4.5 years	When gestation is complete, the first test lactations begin after calving.
5.5 years	First complete heifer lactation *available*

This shows that after 5.5 years, records are available to check the breeding potential or genotype of the dairy bull under test.

This type of test mating is worth far more than picking out a dairy bull on its external appearance or even on its mother's milking performance.

3. There are a lot of examples of inheritance, where dominant genes control characters, which are more easily understood. One of the best examples is that of the colour marking beef bulls. The Aberdeen Angus bull passes on its polled (hornless) character to all its offspring. The calves of a Hereford bull have white marks on their forehead. A Charolais bull passes on its fawn coat colour to all its offspring. Since the bull also passes on a useful potential for liveweight gain, the coat colour helps farmers to identify these young stock at cattle sales.

4. Mendel's theories can also be used to explain hybrid vigour. When two different closely inbred lines of plant or animal are crossed, the offspring can show vigorous growth in the first generation only. This is called hybrid vigour. It is by no means certain that the first generation of such a cross will show hybrid vigour. We have all heard that a mongrel dog is often a healthier and more agreeable companion than a pedigree closely bred animal. To a certain extent this is the result of the same genetical effect.

When animals or plants are closely bred to obtain pure breeding lines (homozygous), dangerous or undesirable recessive genes show themselves in the animals or plants because they are no longer masked by the dominant gene as they were in hybrids with heterozygous genes.

These genes may cause an animal to be susceptible to disease, have weak bone structure, be a poor digester of food, etc.

The offspring of a first cross between two different inbred lines can result in the masking of many undesirable recessive genes and an overall improvement in vigour. Hybrid vigour, therefore, is only likely to occur if,

(a) undesirable recessives were present in the inbred lines;
(b) these recessives are successfully masked by dominant genes from the other animal or plant.

The study of genetics has become an involved science of its own. The scene has recently been complicated by the commercial use of polyploid plants like the tetraploid ryegrasses. These plants have twice the normal set of chromosomes and hence, four genes controlling every character.

Agriculture is the chief outlet for the results of genetical experiments. Keeping up with progress in this direction can be difficult.

This chapter is intended to help the reader to make progress in the right direction.

Section II

The plant

Introduction

To obtain the highest yields from our cash crops and the best nutritive value from our fodder crops for livestock feeding we must prepare the seed bed carefully, use good seed, fertilise the crop adequately and control weeds, pests and diseases.

A knowledge of the structure of the plant is important for identification, and also for appreciating the nutritive value of forage crops.

The physiology of the plant, i.e. how it produces its food, absorbs its water etc., provides information upon which cultural practice is based.

This section covers briefly the structure and function of green crop plants, relating the theory to practice wherever possible.

Green plants can absorb simple chemical compounds from the soil and air to convert into energy-rich carbohydrates and body-building proteins. The plant supplies its own energy requirements for growth and reproduction from its reserves of carbohydrate. It builds its new cells and repairs its old ones from the protein which it is able to make from carbohydrate and nitrates absorbed from the soil.

Animals are dependent upon plants for their supplies of carbohydrate and primary protein. An understanding of plant nutrition can help to achieve maximum production from both plant and livestock enterprises.

Classification of plants and animals

The plant and animal kingdoms are large, including many varied organisms in each. It is not surprising therefore to find that the same plant or animal, appearing in different parts of the world, has been given different names. This leads to confusion and a universally acceptable system of classification is therefore desirable.

Even to a casual observer it soon becomes obvious that those animals with several features in common can be placed in a group. Large groups of animals arise from this primary separation, for example: worms, slugs and snails, fish, reptiles and so on. Within these primary groups smaller divisions can be seen, and even more become apparent on close examination.

Similar groupings can be found in the plant kingdom.

A knowledge of group characteristics may often enable a biologist to classify an animal or plant within a group without knowing its exact name. This ability has great value in agriculture. Suppose a farm animal passes a parasitic worm from its intestine, a rapid examination of the worm would determine whether it was a round or flat worm. There is a different treatment required for each type of worm which can be put into effect to rid the animal of the pest without actually knowing the name of the worm. These groups are very important in the plant kingdom when weeds have to be sprayed with chemical weedkiller.

For example, the removal of certain weeds like wild oats and blackgrass from a cereal crop requires specialist chemicals, since grasses and cereals are in the same plant group. Plants within the same group are often susceptible to the same diseases. Grass weeds like couch grass can perpetuate 'take-all' disease of wheat and barley through a break crop if they are allowed to remain.

The present system of classification for plants and animals is based on the work of 18th century botanists. The binomial system of Linaeus, allots two Latin names to each organism. The first name written with a capital letter, fixes the Genus. This is a group of plants or animals which, although unable to breed together to produce fertile offspring, are very similar in their appearance, structure and habits. The second name, written with a small letter and also in Latin, is the specific name. A species is a group of clearly distinguishable organisms which can successfully inter-breed. Within the species there may be apparent strains or varieties like the different races in the human species.

There are several divisions of the plant kingdom before the genus is reached. As an example, here is a classification of White clover:

Kingdom:	*Plantae*	
Division:	*Spermatophyta*	(Seed bearing plants)
Class:	*Angiospermae*	(Seed plants with covered seed)
Order:	*Leguminosae*	(Plants having a symbiotic relationship with nitrogen fixing bacteria in their roots)
Family:	*Papilionaceae*	(All having sweet pea type flowers)
Genus:	*Trifolium*	(Clover group with 3 leaflets)
Species:	*Trifolium repens*	(*repens* means creeping)

The use of Latin names universally has simplified international recognition of organisms.

Most of the plants of importance in agriculture and forestry are *Spermatophytae*, which includes the *Gymnospermae*, naked seed plants of the coniferous group, as well as the *Angiospermae*, the flowering plants. In this Section the green plants referred to are all flowering plants, with the very odd exception which is specified.

4 Plant morphology and anatomy

(The external and internal structure of plants)

(A) Monocots and dicots

Of the many different types of plant, those which are grown agriculturally are green, have roots, stems, and leaves and produce seeds inside the ovaries of flowers.

These plants are divided into two important groups called monocotyledons and dicotyledons. Monocots and dicots for short.

Monocots, including the grasses and cereals, have one cotyledon in their seeds.

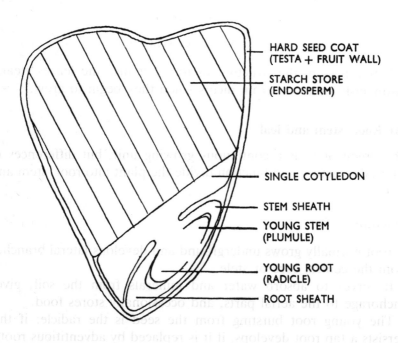

Fig. 4.1 Monocot seed—maize seed section.

Dicots, including root crops and legumes, have two cotyledons in their seeds.

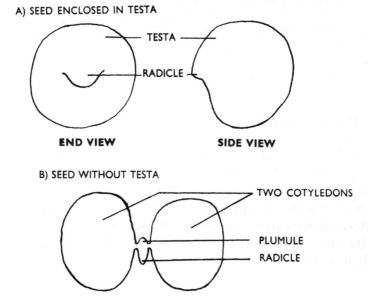

Fig. 4.2   Dicot seed—red clover.

Differences in the structure of roots, stems, and leaves clearly distinguish monocots from dicots when they begin to grow.

(B) Root, stem and leaf

The green plant is a continuous growing unit, but differences in internal structure and function divide the plant into root, stem and leaf.

The root

A root normally grows underground and develops lateral branches from the central core or stele.

It serves to absorb water and minerals from the soil, gives anchorage to the aerial parts, and occasionally stores food.

The young root bursting from the seed is the radicle: if this persists a tap root develops; if it is replaced by adventitious roots, a more fibrous system develops.

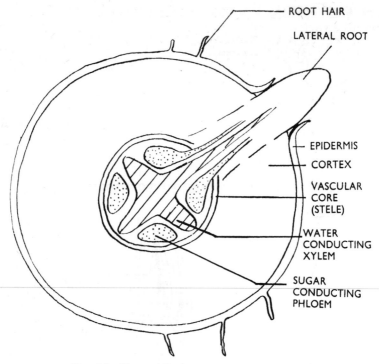

ROOT HAIR

LATERAL ROOT

EPIDERMIS

CORTEX

VASCULAR
CORE
(STELE)

WATER
CONDUCTING
XYLEM

SUGAR
CONDUCTING
PHLOEM

Fig. 4.3 Young dicot root, transverse section.

The young radicle, or any lateral root, grows small root hairs behind its growing tip which absorb water by osmosis and minerals by active selective absorption. The latter process requires energy which is provided by root respiration. Therefore oxygen must be available from the air in the soil.

No soil air means no mineral uptake.

The young absorptive rootlets of a plant are unprotected and readily damaged by soil pests, while older roots are protected by cork bark which prevents them from absorbing nutrients.

The stem

A well formed stem bears flowers well above ground, and spreads its leaves to obtain maximum light and air.

The outer layer (cortex) when young is often photosynthetic, but later on becomes corky. The inner layers (stele) contain conductive

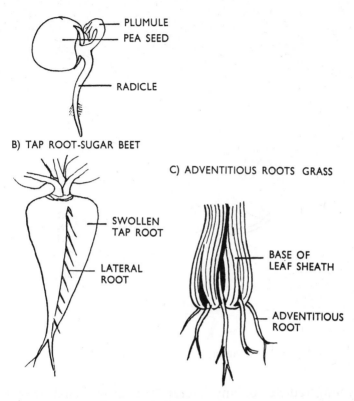

Fig. 4.4 Different types of root.

(vascular) tissue. In very young stems of dicots and all monocots, the vascular tissue is confined to separate threads called vascular bundles. Older dicot stems become secondarily thickened (strengthened) when the vascular tissue forms a complete wide ring which eventually produces the wood of a thick stem.

Vascular tissue consists of xylem, tubular cells carrying water up stems from roots to leaves, and phloem, tubular cells passing sugar solutions about the plant from the leaves to flowers, food stores, roots, etc.

The point at which a leaf is attached to a stem is the node. In the angle (axil) between the leaf and stem axillary buds are formed to produce flowers or branches. The tip of the stem carries the terminal bud which often forms the flowering bud. An actively growing terminal bud usually suppresses the growth of axillary buds below it. To encourage the growth of Brussels sprouts in the autumn many growers remove the terminal bud which would otherwise prevent the sprouts from developing early. Pruning back the terminal bud on woody plants will encourage the plant to develop more lateral shoots (axillary buds) which were being suppressed by the terminal or apical bud.

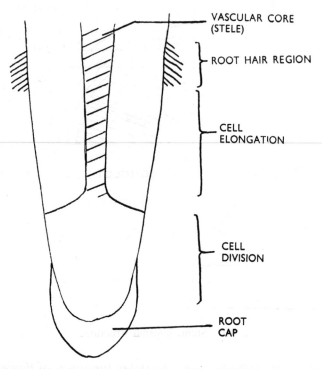

Fig. 4.5 Young absorptive rootlet, longitudinal section.

Dicot stems are usually long, except in root crop plants during their first year. Often the stem trails along the ground, rooting at the nodes.

Monocot stems remain short, like a bulb, until they flower. The short stem is in the vegetative condition; the long ones are called flowering culms. Branches forming on the vegetative stem are called tillers.

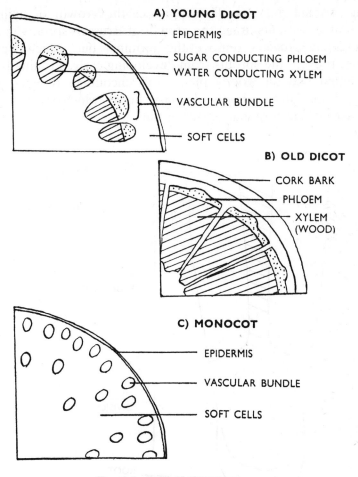

Fig. 4.6 Stem, internal structure.

Removal of the terminal bud can delay the onset of flowering by a short time, but as the year advances and the days grow longer the grass is forced to flower. Flowering increases the fibre content of the grass, rendering it less nutritious to ruminant livestock. The vascular bundles of the monocot stem contain lignin which is an indigestible strengthening fibre.

Monocot stems remain short, live a bulb, until they flower. The short stem is in the vegetative condition; the long ones are called flowering culms. Branches forming on the vegetative stem are called tillers.

The leaf

The leaf is the photosynthetic factory of the plant. The green pigment, chlorophyll, is essential to the process as is an ability to transpire a lot of water through the leaves to create a transpiration to pull drawing water up the stem to the leaves.

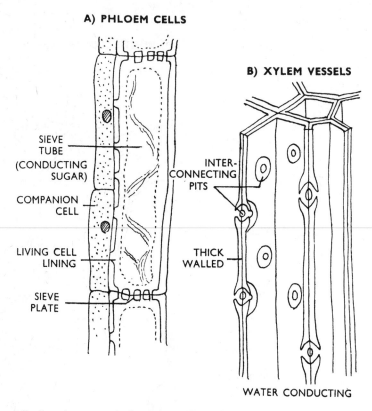

A) PHLOEM CELLS

B) XYLEM VESSELS

SIEVE TUBE (CONDUCTING SUGAR)

COMPANION CELL

LIVING CELL LINING

SIEVE PLATE

INTER-CONNECTING PITS

THICK WALLED

WATER CONDUCTING

Fig. 4.7 Imaginary vertical sections through magnified vascular bundle cells showing phloem and xylem.

The delicate leaf structure is readily damaged by frost, pests, and disease. Herbaceous perennial plants die back to ground level during the winter to protect the whole plant from frost and disease attack through the leaves. Woody plants seal off their leaves during the autumn for the same reason.

Dicot leaves usually have a separate leaf blade (lamina) and leaf stalk (petiole), and grow straight out or at an angle to the stem. They have spreading, branching veins (vascular strands) and frequently have stomata (transpiration and gaseous exchange pores) only in the lower surface of leaves. Dicot leaves are often subdivided into leaflets, e.g. clover has three leaflets on every leaf stalk.

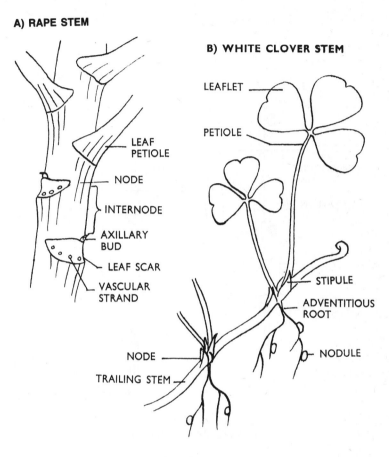

A) RAPE STEM

B) WHITE CLOVER STEM

LEAFLET

PETIOLE

LEAF PETIOLE

NODE

INTERNODE

AXILLARY BUD

LEAF SCAR

VASCULAR STRAND

STIPULE

ADVENTITIOUS ROOT

NODE

TRAILING STEM

NODULE

Fig. 4.8 Dicot stems.

Monocot leaves are not divided into petiole and blade and the base of the leaf wraps around the stem, forming a leaf sheath before it branches away. There is often a distinctive membranous ligule and sometimes small auricles at the junction of leaf sheath and leaf blade. The veins on a monocot leaf are characteristically parallel to one another, and between them there are stomata on the upper surfaces of the leaf.

A) VEGETATIVE STEM

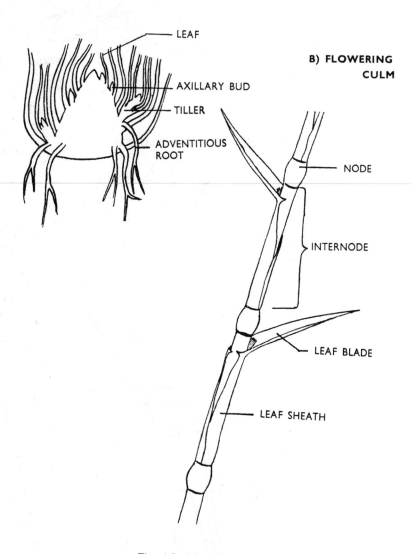

Fig. 4.9   Monocot stem.

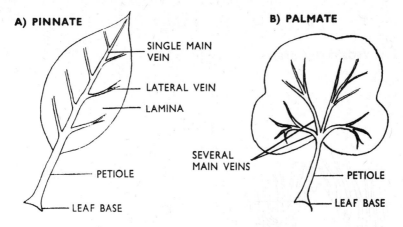

A) PINNATE

SINGLE MAIN VEIN

LATERAL VEIN

LAMINA

PETIOLE

LEAF BASE

B) PALMATE

SEVERAL MAIN VEINS

PETIOLE

LEAF BASE

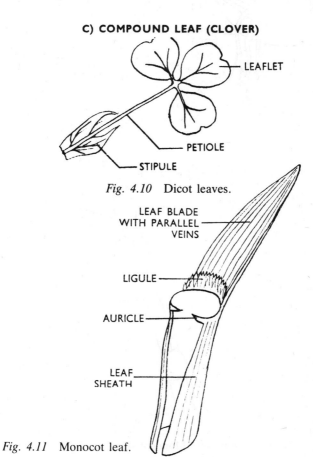

C) COMPOUND LEAF (CLOVER)

LEAFLET

PETIOLE

STIPULE

Fig. 4.10 Dicot leaves.

LEAF BLADE WITH PARALLEL VEINS

LIGULE

AURICLE

LEAF SHEATH

Fig. 4.11 Monocot leaf.

A) SECTION THROUGH LEAF

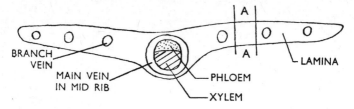

B) ENLARGED SECTION THROUGH 'AA' ABOVE

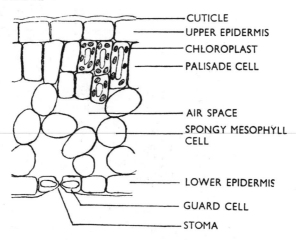

Fig. 4.12 Dicot leaves.

When studied in microscopic section the delicate structure of a leaf can be seen to fit its function very well.

It must allow light to penetrate, so it is thin. To allow air to enter and water to escape, the holes in the epidermis, called stomata, can open under the control of guard cells. Many air pockets interconnect within the leaf allowing free air circulation. The green pigment chlorophyll is concentrated into chloroplasts which are gathered in the active cells.

The species, or groups of plants which can interbreed, have many morphological characters in common. These characters help us to distinguish one species from another in the field.

5 Plant physiology

(How a plant functions)

Energy can neither be created nor destroyed, but it constantly changes from one form into another.

Livestock obviously require energy for movement and for keeping themselves warm. Plants also need energy to carry out their unseen functions, such as absorbing nutrients and transporting food solutions. All this energy comes primarily from the sun. Through the activity of the green plant this light energy is held in chemical form supplying the needs of both plant and animal.

(A) Photosynthesis

This is the build-up of chemical compounds of carbon, hydrogen and oxygen. The final products have a high energy content and are called carbohydrates.

Carbohydrates are used in respiration of both plants and animals to supply energy for cell working. Carbohydrate is converted sometimes to fat, an even higher energy container, for compact storage. Carbohydrate is also used with nitrogen to form protein for body and plant cell building.

Only plants containing the green pigment chlorophyll can photosynthesise. Usually the process only takes place in leaves.

The energy required for this endothermic chemical reaction comes from sunlight. Leaves must be thin and delicate to allow light to penetrate to the middle leaf layers (mesophyll) which contain most chlorophyll.

Factors essential for photosynthesis

Chemically, photosynthesis is described by this equation:

$$nCO_2 + nH_2O \xrightarrow[\text{chlorophyll}]{\text{sunlight}} (CH_2O)_n + nO_2$$

Carbon   Water   Carbohydrate   Oxygen
dioxide

This can only occur in a living plant cell in the presence of the catalyst chlorophyll.

A catalyst speeds up chemical activity.

Carbon dioxide, water, sunlight, and chlorophyll will therefore be essential to the process and a suitable temperature is also desirable.

Carbon dioxide is present in small amounts (0.03 per cent) in the air which enters the leaf through the stomata. This is often a limiting factor in photosynthesis, as shown by glasshouse experiments. The organic matter content of soil is slowly decomposed to release carbon dioxide around plants.

Water enters a plant through root hairs by osmosis and is forced up the stem to the leaf largely by transpiration pull.

Phosphate is essential for root development. Pests and disease damaging roots or stems can reduce the water supply to the leaves.

High relative humidity around the leaf reduces transpiration pull. It is therefore important to thin the crop and remove weeds early, to prevent the atmosphere becoming too moist around the plant. Weeds will also compete with the crop at root level for available water. Irrigation can increase the water content of soil and organic matter in the soil can conserve water for plants.

Chlorophyll is the organic catalyst which is essential to the reaction. The greater the leaf area containing this green pigment, the better is the possible production of carbohydrate.

Nitrate is the fertiliser to encourage leaf growth, and magnesium and iron are needed to make chlorophyll. The absorption of these minerals will be affected by the oxygen content of the soil; waterlogged soil contains no oxygen. Pests and diseases of leaves can rapidly reduce effective area. Viruses, bacteria, fungi, eelworms, and insect pests, reduce available green leaf and cut off the nutrient supply for photosynthesis.

. The intensity and duration of sunlight is very important. A plant which has developed its maximum leaf area in time to benefit from the long warm summer days gives a high yield of carbohydrate. September sun can never compensate for a lack of sun in June.

Shading of plants can be kept to a minimum by trimming hedges, controlling seed rate, singling crop plants early, and weeding by cultivation or use of chemicals.

Temperature of the air affects the rate of photosynthesis which in winter takes place very slowly and only in those plants which retain their leaves. In spring the rate speeds up. From $0°C$ to $35°C$, every $10°C$ rise in temperature doubles the potential rate of photosynthesis.

The elaboration of carbohydrates, fats and proteins

The primary products of photosynthesis are probably three-carbon compounds which are rapidly built up into simple sugars for translocation about the plant. These sugars can be reconverted to three-carbon compounds by the action of plant enzymes, and then by removing water molecules, fats are built up. Fats contain more chemical energy than carbohydrates, they are therefore an economic form in which to store food.

From the three-carbon compounds proteins can be made by green plants able to obtain adequate supplies of nitrate. The nitrate (NO_3) has to be reduced to the amino group (NH_2) before it is attached to the carboxyl $(C = O)$ or acidic groups of the three-carbon compounds.

The whole of the animal kingdom is dependent upon the green plant for its production of protein. Although animals have the ability to change the form of protein molecules, they are unable to carry out the primary synthesis.

Translocation of elaborated sugars etc.

The green leaves of a plant are cleared of their photosynthetic products during the night. The pathway of the sugars appears to be through the stem, in specialised conducting cells called phloem. The phloem cells are elongated and tubular. They contain cell sap and are lined by active cytoplasm. It is thought that the movement of this cytoplasm may assist the movement of organic compounds for the plant. Each phloem cell is associated with a companion cell which probably assists the former to function.

The destination of translocated carbohydrates depends on the structure of each plant. Potatoes obviously transfer carbohydrates to their underground tubers where insoluble starch is precipitated. Sugar beet stores sugar in its root as a winter supply of energy food.

(B) Water absorption and movement in the plant

Water absorption

Most of the crop plants grown in the British Isles are herbaceous. That means that they have very little woody tissue in their stems and rely upon the rigidity of fully water-inflated cells for their support. A fully-inflated cell is called turgid and a deflated one flaccid.

Water is also necessary for photosynthesis, but the plant will have wilted long before water becomes a limiting factor.

Water enters the plant through its roots by way of its root hairs. (See Figs 4.3 and 4.5.)

The root hairs penetrate the soil contacting water and minerals adsorbed on the surface of soil particles.

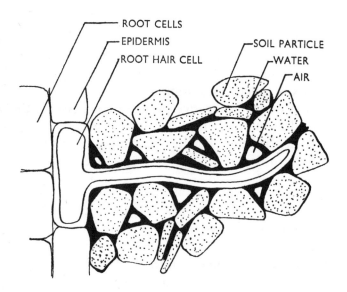

Fig. 5.1 Root hair between soil particles.

Osmosis causes the water to enter the root hair. Osmosis is the passage of water from a dilute solution into a more concentrated solution through a semi-permeable membrane. The semi-permeable membrane is the living cytoplasm lining the cell wall. The concentrated solution is formed by sugars dissolved in the cell sap.

Suction pressure

A suction pressure develops in the root hair cell as a result of osmosis

$$\text{Suction Pressure} = \text{Osmosis} - \underset{\text{(Resistance of cell wall)}}{\text{Turgor Pressure}}$$

i.e. when sucking power of the root hair is equalled by the resistance of the cell wall to further stretching, no more water enters because the cell is turgid.

The combined suction pressure of the root cells can force water a short way up the stem by root pressure.

Water movement (Translocation)

Water travels through the stem, petiole, midrib, and branch veins in the xylem vessels of the vascular bundles. (See Figs 4.6 and 4.7.)

The leaf loses water to the atmosphere through the stomata. This water loss is called transpiration.

Transpiration exerts a negative pressure (suction) on the water in the xylem, drawing it up the stem to the leaf.

Guard cells control the opening and closing of the stomata and these only open in daylight.

Transpiration rate

Transpiration rate is governed by:

a) The number of stomata on the leaf.
b) The temperature of the leaf surface.
c) The humidity of the atmosphere.

In very dry weather when the humidity of the air is low, wilting occurs, i.e. the plant droops. This is caused by the rate of transpiration exceeding the rate of water absorption by the root.

Grass leaves of red fescue roll inwards so that their stomata open into a tube which rapidly becomes very humid. This is one way in which plants have become adapted to combat drought.

Transplanting is best carried out in moist humid weather since the plant loses its root hairs when it is lifted from the soil. In humid weather the plant will not need so much water and will have an opportunity to grow fresh root hairs.

A) SECTION THROUGH AN OPEN
FESCUE LEAF

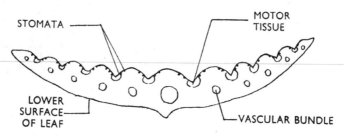

B) SECTION THROUGH A FESCUE
LEAF ROLLED INWARDS

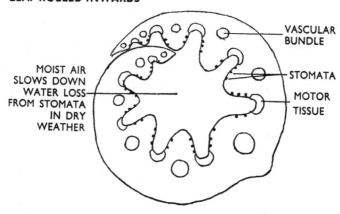

Fig. 5.2

(C) Mineral absorption

The absorption of minerals by the root appears to be partially selective. It can be observed that without adequate oxygen in the soil, minerals are not taken into the plant. This indicates that respiration must take place in the root to provide energy necessary for absorption.

Preparation of a seed bed incorporates air into the soil which will be valuable for mineral uptake.

A) DIAGRAM OF PLOUGHED SOIL

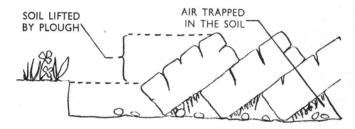

B) DIAGRAM OF SEED BED

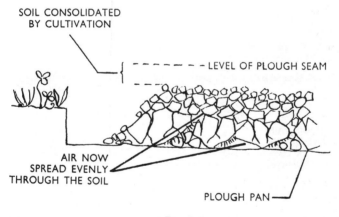

Fig. 5.3

Some minerals are required by plants in regular quantities. These are the major nutrients.

Other minerals are only required in small amounts. These are known as trace elements.

The major nutrients

Nitrogen (nitrate (NO_3^-) increases tillering, leaf growth, photosynthesis, and yield. Legumes (clover, peas, beans, etc.) have bacteria in the nodules on their roots which convert atmospheric nitrogen into a form which is available to the clover plant and also to crops grown in conjunction and afterwards in the rotation.

Phosphorus (phosphate $PO_4^=$), encourages root establishment and is particularly important to legumes. Phosphate also encourages the plant to reach maturity and flower.

Potassium (K^+) is necessary for the movement and storage of carbohydrates in plants. It is particularly important to sugar beet, to legumes to improve phosphate utilisation, and for grass productivity where heavy grazing is practised.

Magnesium (Mg^{++}) is required by a plant for chlorophyll formation.

Iron (Fe^{+++}) is also concerned with chlorophyll formation.

Calcium (Ca^{++}), is used in the formation of cell walls so is vital for plant growth. Acids which form within the plant are neutralised by calcium.

Sulphur ($SO_4^=$ sulphate), is found in some of the vital protein molecules formed in plants.

Trace elements

Boron (BO_3^- Borate); its absence causes 'heart rot' in sugar beet and 'brown heart' in swedes.

Manganese (Mn^{++}); its absence causes speckled yellows in sugar beet and mangolds.

Zinc (Zn^{++}) and Copper (Cu^{++}) are only required in small amounts for normal plant functioning.

Molybdenum (Mo^{++}) is very useful for the development of nodule bacteria in the legumes. Plants occasionally absorb this mineral to excess of requirements. The plant does not suffer but ruminant livestock feeding on the plants suffer from molybdenumosis.

Minerals absorbed as ions

Only chemical compounds which are soluble in water can be

absorbed by plants since the plant absorbs ions. When most compounds are dissolved in water they are separated into their constituent ions.

An ion is an electrically charged particle. Some ions are positively charged and called cations (+) while others are negatively charged and called anions (−).

All the minerals named above have the symbol and sign of the appropriate ion following.

The dissociation of sodium chloride in water would be as follows:

$$\text{sodium chloride (NaCl)} \; \underset{\leftarrow}{\overset{\rightarrow}{} } \; \text{sodium (Na}^+\text{) and chloride (Cl}^-\text{)}$$

Clay particles in the soil carry negative charges. The cations (+) from the minerals are attracted to and held by the clay against leaching. The anion, for instance the valuable nitrate anion, is not held by clay so is easily leached from the soil in drainage water. A high concentration of ions in the soil around a root can prevent it from absorbing water properly. For this reason, sowing the seed after broadcasting the fertiliser may give a seedling a better start than if it were dropped into close contact with the fertiliser.

A) SEED DRILLED AFTER BROADCASTING FERTILISER (SEED NOT IN CLOSE CONTACT WITH FERTILISER)

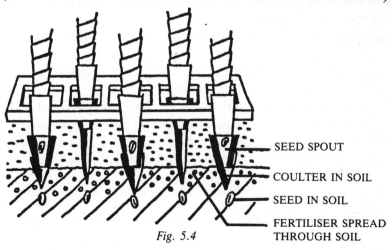

SEED SPOUT

COULTER IN SOIL

SEED IN SOIL

FERTILISER SPREAD
THROUGH SOIL

Fig. 5.4

B) COMBINE DRILLING
(SEED AND FERTILISER TOGETHER)

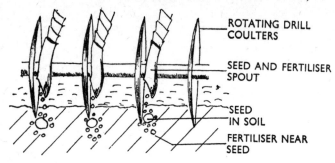

ROTATING DRILL COULTERS

SEED AND FERTILISER SPOUT

SEED IN SOIL

FERTILISER NEAR SEED

Fig. 5.4 cont.

6 Plant propagation

Propagation is the increase in numbers of a species of plant. The two important means are:

a) vegetative
b) seed

In vegetative propagation part of the parent plant breaks off to form the daughter plant.

In seed propagation the fusion of a male and female gamete precedes the formation of a seed inside the fruit.

(A) Vegetative propagation

Parts of plants become detached from the mother plants and develop root, stem, and leaves of their own.

This occurs naturally by stolons and tubers in potato; rhizomes in couch grass; and runners in creeping buttercup.

Artificially it can be helped by taking cuttings, splitting up bulbs, and transplanting tubers.

The advantages of vegetative propagation are that plants establish quickly and all daughter plants are identical to the parent. This is important to a grower who is building up stocks of 'seed' potato, and who is anxious to retain all the characters of the variety in all the tubers which he sells.

The disadvantage of vegetative propagation is that any disease developed by the parent plant can be passed into the daughter plant. Example:

Virus leaf roll or virus mosaic of potato is passed from parent plant into the tubers. If these infected tubers are used for transplanting, they could quickly cause the infection of a large number of plants.

Vegetative propagation is of no use to the plant breeder who is looking for new varieties.

A) CREEPING BENT RUNNER

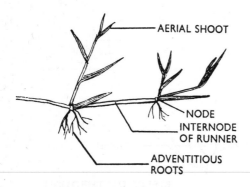

B) COUCH GRASS RHIZOME

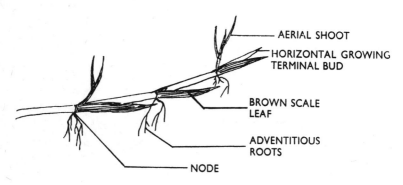

Fig. 6.1   Vegetative propagation.

(B) Seed propagation

A collection of flowers on a stem is called an inflorescence.

A flower is a collection of modified leaves bearing spores which will give rise to gametes.

The important parts of a flower are the stamens and carpels. The stamens form pollen which produces the male gamete. Carpels contain ovules which produce the female gamete.

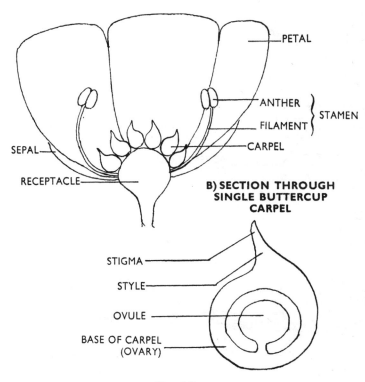

A) BUTTERCUP FLOWER IN SECTION

PETAL

ANTHER }
FILAMENT } STAMEN

CARPEL

SEPAL

RECEPTACLE

B) SECTION THROUGH SINGLE BUTTERCUP CARPEL

STIGMA

STYLE

OVULE

BASE OF CARPEL (OVARY)

Fig. 6.2

Pollination

This is the transfer of pollen from stamen to carpel. Most flowers are bisexual, but the pollen is usually transferred to carpels on another flower.

(i) Cross pollination

This is the name given to the transfer of pollen from the stamens of one flower to the stigma (receptive part of carpel) of another flower of the same species. The transfer is effected largely by insect or wind.

Usually flowers pollinated by insects have brightly coloured petals to attract insects. Scent is also attractive to insects particularly moths which fly by night. Nectar is produced which encourages insects to enter the flowers.

Wind pollinated flowers have no attractive petals. Their stamens have large floppy anthers which blow about. Vast amounts of pollen are produced and the stigma on the carpel is feathery to catch the blown pollen.

The grass flower is an important example of wind pollination.

There are two forms of inflorescence found in grasses: the panicle which is branched, bearing spikelets at the ends of the branches; and the spike which is unbranched with spikelets born directly on the central stem or axis.

Usually two outer glumes protect two or more florets in each spikelet.

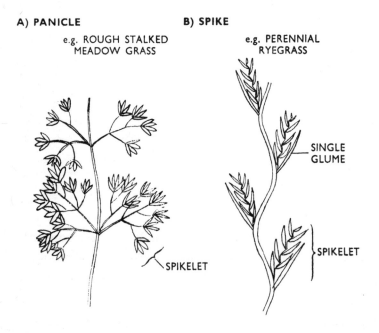

Fig. 6.3 Grass inflorescences.

Each floret has at its base two leaves, the lemma or outer pale and the palea or inner pale. The lemma often has a bristle or awn on its back. Inside are three stamens and a carpel on which are two feathery styles and stigmas.

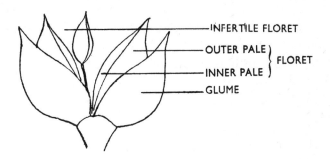

DIAGRAM OF SPIKELET SHOWING RACHILLA

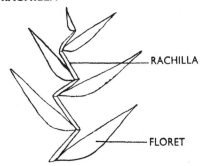

Fig. 6.4   General view of spikelet.

(ii) Self pollination

Weeds often have devices which enable them to pollinate themselves with pollen from the same flower. This means that seed will be produced from a single plant growing in isolation.

Fertilisation

At fertilisation, a gamete formed from the pollen fuses with a gamete formed by the ovule.

The pollen grain grows a tube from the stigma into the base of the carpel where the ovules are. Gametes are released from the pollen tube to fuse with gametes produced within the ovule.

Seed and fruit development

The fused gametes (zygote) develop to form the embryo within the seed. The ovule forms the rest of the seed, i.e. the food store or endosperm and the seed coat or testa. The carpel now enlarges to form the fruit.

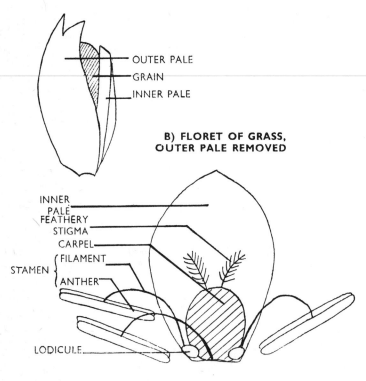

A) FLORET OF WHEAT WITH RIPE GRAIN

OUTER PALE
GRAIN
INNER PALE

B) FLORET OF GRASS, OUTER PALE REMOVED

INNER PALE
FEATHERY STIGMA
CARPEL
STAMEN { FILAMENT
ANTHER
LODICULE

Fig. 6.5

Note that no fruit or seeds are formed unless fertilisation takes place. A good blossom on a pear tree does not automatically mean there will be a lot of fruit.

Seed dispersal

The seed is dispersed, sometimes still within the fruit, naturally or by the farmer.

The seed of a dicotyledon has two seed leaves (cotyledons) which usually contain the food store (starch) for germination. (See Fig. 4.2.)

The cotyledons are attached to the young root (radicle) and the young stem (plumule). All this is surrounded by the seed coat (testa) in which there is a small hole (micropyle) for water absorption.

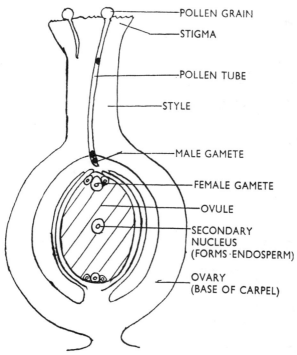

Fig. 6.6 Carpel at fertilisation.

The monocotyledon seed has a large store of starch in the endosperm. (See Fig. 4.1.) The single cotyledon (scutellum) absorbs the endosperm and passes it to the root and young stem, each of which is enclosed in a protective sheath. Many monocot seeds are released from the plant whilst still encased in their flowers. Most grass seeds, oats, and barley are like this. The advantages of seed propagation are that daughter plants differ slightly from the characters of the parent plants; a fact which is made use of by the plant breeder.

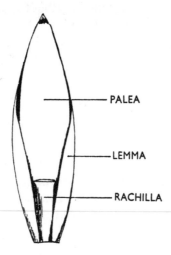

Fig. 6.7   Grass seed, perennial ryegrass.

Seed plants are usually vigorous in their early life but can have difficulty until they are established.

(C) Germination

The seed of the plant is dormant until water, oxygen, and warmth are supplied together. Some seeds also require a short time at a low temperature before dormancy is broken. Breaking dormancy is germination.

As germination commences, the seed absorbs water through the micropyle and more slowly through the testa. When the enzyme diastase is thoroughly wet, it is activated and rapidly converts starch to sugars as the temperature rises.

The testa is split by the swollen damp seed and, as oxygen reaches the embryo, respiration increases and energy for growth is produced in abundance. The radicle now grows to absorb water, and the plumule follows later breaking through the soil surface to begin photosynthesis.

The temperature required for germination in the soil varies from plant to plant. Rye, a hardy plant, will germinate at 2.2°C, which accounts for its use as a cereal crop in colder climates where it is essential to begin growth as early as possible because of the short growing season. Wheat, barley, and sugar beet germinate slowly at temperatures above 4.4°C. Maize will only begin growth at 8.8°C. Hence maize is difficult to harvest in Britain because our growth season is too short.

Types of germination

(i) Epigeal germination

The stalk between the radicle and cotyledons (hypocotyl) grows, pushing cotyledons out of the ground to become photosynthetic. From these, the plumule grows to form the first true leaf. Examples are found in the genus *Brassica* and the order chenopodiaceae (sugar beet and mangold).

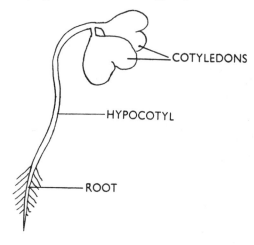

Fig. 6.8 Epigeal germination, kale.

(ii) Hypogeal germination

The plumule grows above ground leaving the cotyledons in the soil. The first green leaves to appear are true leaves. Examples: field beans and all monocots.

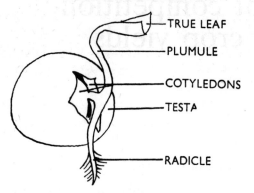

Fig. 6.9 Hypogeal germination, dicot, pea.

7 Plant competition and crop yields

(A) The maximisation of crop yields

Photosynthesis, i.e. carbon assimilation in green crop plants, only occurs during the day-time, whereas respiration takes place within the plant all the time.

Net assimilation rate, carbon assimilation minus respiration loss, is therefore a better measure of a plant's growth rate. Net assimilation rate is controlled by environmental factors like light intensity, air temperature, chlorophyll content of the leaf and competition for carbon dioxide, water and nutrients.

Crop yield is determined by the leaf area available for photosynthesis multiplied by the net assimilation rate. To maximise yields of a crop plant it is essential to obtain complete ground cover by plant leaves as early as possible after planting, thereby avoiding any wastage of light which would otherwise fall on bare ground.

Not enough leaf will result in yield reduction, too much leaf causes shading. There is an optimum leaf area in relation to ground covered for each plant. Leaf area index is this relationship between leaf area and soil surface.

$$\text{Leaf area index} = \frac{\text{Leaf area}}{\text{Ground area}}$$

If leaf area index is less than 1 then light is being wasted, because some of it must be falling on bare ground.

The optimum leaf area index for grasses appears to be around 7 or 8, for sugar beet 4–5, and for potatoes 3.

To achieve the highest crop yields, the optimum leaf area should be attained if possible before the season of maximum light intensity which coincides usually with the longest day. For example, delays in planting main crop potatoes after the end of the third week of April, result in reductions of tuber yield, because the optimum leaf area has not been reached, by the time when the daylight is at its best for photosynthesis. These yield reductions are in the order of 125 kg per ha per day delayed at planting, or 900–1,800 kg per ha per week.

Inter-row spacing and seed rate are therefore very important factors as they will govern the speed at which the crop covers the ground in its early stages of growth.

(B) Requirements for healthy growth

If a green plant is to achieve its maximum growth rate it must have an adequate supply of water, sunlight, fresh air, minerals, and warmth. If plants are growing too close together they compete with one another for light, water, minerals and air.

The plant population is very important if a farmer is to achieve maximum productivity per acre. Too many plants may result in a crop of small plants of low quality and a high incidence of disease, whereas too few could cause large woody unsaleable or inedible plants to develop.

Seed rates are carefully adjusted to reduce competition between plants and, where necessary, crop plants should be singled and spaced out at an early stage of growth before they compete strongly.

Competition also arises from weeds. A weed is a plant growing out of place, e.g. a potato which comes up in a cereal crop is a weed. Those weeds which present the greatest problem are usually wild plants which have developed efficient methods of propagation by seed or vegetative means.

(C) Weed control

The best controller of weeds is a healthy growing crop. Once the leaves of the crop plant cover the soil surface there is no longer any problem with weeds.

Mechanical

Weed control by mechanical means aims to kill weeds in one or more of the following ways:

1. Removing by dragging implements through the soil; followed by gathering weeds into heaps and burning them.
2. Disturbing them constantly to prevent the weed from replenishing its food reserves. This often requires the soil to be left without a crop for part or all the season which is, of course, very expensive. Sometimes cultivation encourages weeds to germinate, then disturbance of the seedlings can be fatal.
3. Burying the weed deep enough in the ground often kills it, but this will not work with all weeds.
4. Chopping the weed into very small pieces with a rotary cultivator often finishes their chance of survival.

Every mechanical operation is costly, shortens the time available for crop growth, and removes valuable moisture from the soil.

The weed problem is an annual event, as the hedgerow will re-infect fields and seed may blow in from neighbouring land.

Chemical herbicides

Complete herbicides will clear the soil of all plants and some are persistent in the soil for a long time, such as sodium chlorate (up to six months), or have a very short effective life, like paraquat and diquat (possibly less than a day).

Selective herbicides used at the right concentration will affect certain plants more than others. The difficulty arises when the weed is very similar, even related, to the crop plant. Under these circumstances it is difficult, often impossible, to select weed from crop.

The most common selective weedkillers in use at the present time are of the growth regulator type. Their effect on plants is to force them to outgrow their reserves of strength. Although frequently not fatal to the weed, they cause a set-back in growth which allows the crop plant to dominate the field.

Most of these weedkillers enter the plant through the leaves, although some enter through underground rhizomes and the poison is then translocated about the plant in the vascular bundles.

Before using a herbicide it is advisable to be certain that it will not damage the crop plant. In many cases late spraying of cereals damages the developing ear. Remember also that the wind can carry spray on to adjoining susceptible crop plants; or could cause a double dose of spray to fall on the crop, causing damage.

Effectiveness of the herbicide will depend on the stage of growth of the weed and its susceptibility to the spray.

Persistent use of one type of spray chemical on a field can encourage the development of weeds which are resistant. Many of the herbicides now used can best be described as 'cocktails', i.e. a mixture of several chemicals to control a broad spectrum of weeds.

Section III

The animal

Introduction

In the same way as understanding of the soil and plant is necessary to enable an arable farmer to use present-day scientific advances to produce high yielding crops of good quality, so is knowledge of the construction and working of the animal body useful to the stock farmer. On considering any stock enterprise, it will be realised that profits depend either on the efficiency with which animals (or birds) convert food to animal products, or on the successful breeding and reproduction of livestock. Emphasis will, therefore, be placed on the study of the digestive and reproductive systems of animals: some consideration, however, of other structures will be necessary to understand fully the working of the animal body.

8 Anatomy

(A) Basic structure

The animal is a complex living processing factory and depends on the activities of the living cells of which it consists.

On examining parts of the animal, concentrations of large numbers of cells, specialised in a particular function, are found. In the muscles of the animal are collections of cells able to change shape: in other places there are cells which have developed the activity of chemical synthesis. Cells of similar specialised activity form tissues so it is possible to identify muscular tissue, epithelial (lining) tissue, and glandular tissue. Where a tissue performs a particular function, such as moving a limb or producing milk, it is organised into a functional unit known as an organ, e.g. a muscle or an udder. Where various organs are related in their activity, possibly performing complementary parts of a complex process, they collectively become a system such as the nervous system or the respiratory system. Therefore, a body function is carried out by a system of organs, each organ being made up of tissues consisting of masses of similar cells. This complicated 'works' of an animal functions most efficiently in an environment which is liquid-supported, warm, and protected from other harmful living organisms. Study must here be made of the means by which this environment is provided.

Body structure

In the first instance it is necessary to think of an animal as a functioning mass of jelly-like shapeless cells, and then to consider how this mass might be organised into an animal. There are really two possible ways to construct a living animal: either all the parts could be packed into a container or they could be fixed onto a frame. The 'put the parts in a box' technique is seen in the construction of an insect: the hard exo-skeleton contains all the soft structures and successfully protects them. Such a method of construction has disadvantages: limb attachment is difficult and growth is a problem, but the greatest difficulty is the weight of the 'box' if the animal is of any size.

In larger animals, the alternative construction method is found: the body parts are fixed to a frame. In this way the problems of the previous method are overcome, but protection is not so good. The main supporting member of the frame is the backbone; this is rigid enough, with the aid of muscles, to take the weight and yet it is sufficiently flexible to avoid clumsy movement. The internal organs are either fixed to the muscle layers surrounding the backbone or hung on sheets of membrane from this structure. To achieve movement, the whole body is supported on limbs with flexible joints controlled by muscles. The very important tissues of the body are surrounded by bone to provide protection—as in the case of the brain inside the skull—while the overall protection is given by the skin which also forms a bag to contain the liquid environment in which the body cells exist.

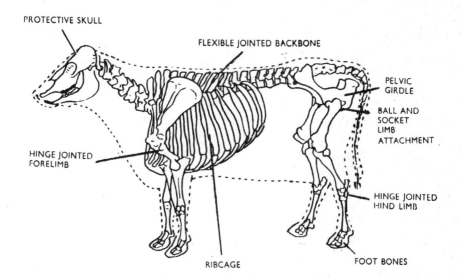

Fig. 8.1 Animal skeletal frame.

Before considering the systems of the body, both the joints of the framework and the functions of the skin must be studied because of their injury possibilities.

(B) Joints and bones

The limbs, formed by 'long' bones, enable the animal to walk because the bones are joined by hinge joints. A hinge joint is very similar to a bearing in a machine where the same problems have to be overcome: one part has to rest on, but move relative to, another part.

The ends of the bones are enlarged to increase the area in contact, and these bearing surfaces are covered with softer cartilage—the bone ends often being shaped to fit together. The bones are kept in alignment, and the joint sealed, by a ligament made of flexible elastic tissue. To avoid any friction in the joint, it is lubricated by a natural oil, synovial fluid.

If, on injury, the ligament becomes perforated and the lubricant is lost, then pain is experienced and wearing away of the bones occurs. An arthritic condition is caused if small crystals are formed in the lubricant, and damage and pain will result. Occasionally, injury can cause the two contacting bone surfaces to grow and fuse together; the result of this is the loss of the 'hinge' power of the joint.

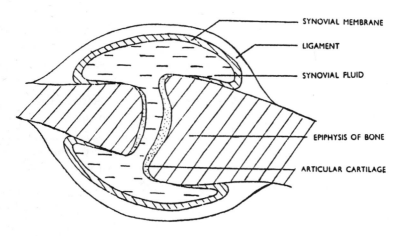

Fig. 8.2 Simple bone joint.

The comments above about bones growing, give rise to the question: what is bone? Bone, as it is most commonly seen, is dead and the impression therefore gained is that it is a white, hard, lifeless material. However, if a microscopic examination of bone is made, it is seen to consist of a network or mesh of living cells embedded in white calcium phosphate.

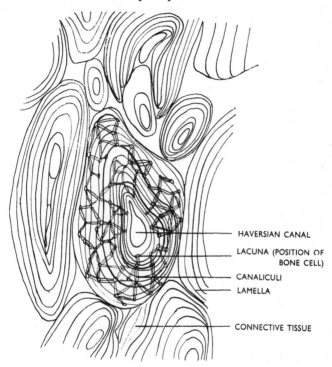

HAVERSIAN CANAL

LACUNA (POSITION OF BONE CELL)

CANALICULI

LAMELLA

CONNECTIVE TISSUE

Fig. 8.3 Transverse section of bone seen under the microscope.

Bone cells appear to be very static, but radioactive labelling techniques have recently shown that new growth continually occurs on the outer surface of long bones as the result of migration of cells from the inner surface (the bone is hollow) to the outer surface. In one day the bone cells are able to form matrix (white calcium phosphate) equal to their own weight.

(C) Skin

As already mentioned, the skin is a waterproof bag maintaining a liquid environment in which the living body cells function. If the skin is grazed, the liquid present in the body tissues will seep out. But, in addition, the skin has other important functions. It provides a barrier to organisms such as harmful bacteria which, should they once gain entry, could multiply very rapidly in the ideal conditions existing inside an animal. An animal's surface suffers friction with its surroundings, and one of the skin's functions is to produce dispensable material on the surface in the form of dead skin cells, which can be harmlessly worn away.

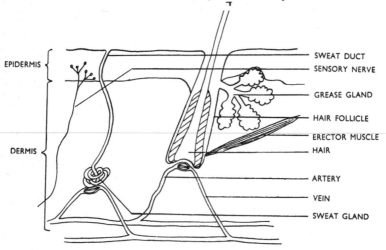

Fig. 8.4   Vertical section through skin of mammal.

In addition to these rather mechanical functions, the skin has an important part to play in heat regulation. An essential part of the skin structure is the layer of fat which is a heat insulator having a constant heat retention function. In addition, the animal body has two ways of varying heat loss from the skin. The size of the tiny blood vessels in the skin can be varied and, therefore, it is possible to control the supply of blood which will be cooled by passing near the surface. The coat of the animal consists of hair which, by movement of the tiny muscles attached to the root of each hair, can be increased or decreased in thickness—another variable heat control feature of the skin.

From these facts it will be obvious that a young animal, without much thickness of fat in the skin, will need warmer housing conditions than a more mature animal. Similarly, animals such as pigs, with a poor coat development, will suffer from temperature variations much more than will a thick-coated animal.

9 The digestive system

All the processes of animal production are dependent on the animal being presented with suitable food and the animal's ability to extract from that food the raw materials which it requires.

The feeding stuffs fed to farm animals consist of mixtures of complex organic compounds, simple chemical salts and water. The organic compounds have been formed in other living cells, such as the starch formed in green leaves. Organic compounds consisting of atoms of carbon, hydrogen and oxygen built into molecules are known as carbohydrates. These are energy-containing foods which may be complex and difficult to break down, such as the cellulose in straw, or they may be simple in chemical form and easily dealt with in the animal's digestive process. Fats and oils formed from the same chemical constituents are more concentrated forms of energy foods.

Proteins are the most important group of chemical compounds in food and are formed of chains of molecules each built from carbon, hydrogen, oxygen and nitrogen atoms. The individual links in these chains are amino-acids of which more than twenty commonly occur, their numbers and arrangements producing thousands of different proteins. To build animal cells eleven of these amino-acids are essential and can only be derived from ready-made proteins, so an animal's diet must contain such protein if it is to grow.

Food also contains inorganic chemical compounds known as minerals which supply some of the essential elements for the formation of the chemical reagents used in the animal body. For example, iron is necessary for the blood protein haemoglobin, the red respiratory pigment; calcium is essential to bone formation; iodine is a constituent of the growth hormone thyroxin. A deficiency of any one of these minerals, even though the amount required is minute, can seriously affect the efficiency of the body as a whole.

Most chemical reactions carried out in the laboratory in a test-tube, need water to be present, in fact the chemicals often have to be dissolved in water for any activity to occur, and the same is true for reactions in the animal body. In addition protoplasm, the living constituent of cells, contains about 70 per cent water. These wants are supplied by the water which an animal drinks and the moisture taken in in food.

For these food constituents to be able to enter the body, they must be broken down to chemically simple units which can be absorbed into and through the layers of cells which line the digestive tract. This breakdown process is digestion. The chemical actions on large masses of food would be slow and inefficient and therefore an essential feature of digestion is mechanical fragmentation of the food to increase the surface area over which chemical action can occur.

The amount of nutritional value which is available to the animal from the food and the quantity which will pass through the system unused, will depend on the efficiency of these related processes of digestion and assimilation. When considering any sample of a food, the proportion of the total content of feeding value used by the animal is expressed as the percentage of digestibility.

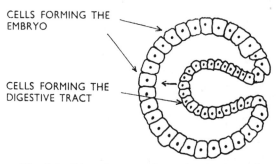

CELLS FORMING THE
EMBRYO

CELLS FORMING THE
DIGESTIVE TRACT

Fig. 9.1 Digestive tract formation in the embryo.

An animal is said to eat food which enters the body: there is, however, a sense in which during digestion this food is still outside the body system. During the embryonic development of an animal, while it is still a ball of cells, the surface becomes 'pushed in' at one point and the indentation progressively grows through the centre of the cell ball until it meets the outer cell layer on the opposite side. The two cell layers then join and perforate to form a through channel which eventually becomes the digestive tract—or alimentary canal. The outside has, therefore, been turned inside. For material to be truly inside the animal body system, it must pass from the digestive tract through the wall: this process is called assimilation.

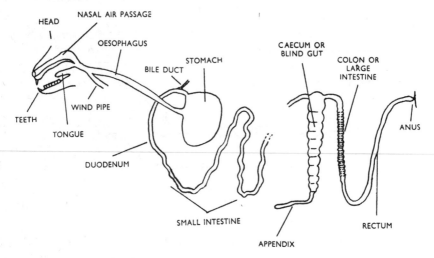

Fig. 9.2 The organs of the digestive system.

To understand the details of digestion it is necessary to study each of the organs which make up the system.

(A) The mouth and the transport of food

The feel of food by the lips and the taste registered by the sensitive taste buds on the tongue enable the animal to assess the food's palatability and to select what it consumes. The teeth are used to cut the food into small pieces which are then ground up and pulped by the grinding teeth—the molars. Saliva is secreted from glands at the base of the tongue and mixed with the food; this secretion acts as a lubricant which assists the passage of the food along the gullet, or oesophagus, to the stomach. To enable the digestive system to move food continually, it is necessary for the food material to be in

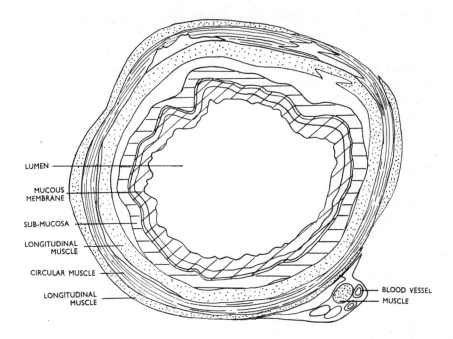

LUMEN

MUCOUS
MEMBRANE

SUB-MUCOSA

LONGITUDINAL
MUSCLE

CIRCULAR MUSCLE

LONGITUDINAL
MUSCLE

BLOOD VESSEL

MUSCLE

Fig. 9.3 Transverse section through the oesophagus of rabbit, approximately 10 mm in diameter.

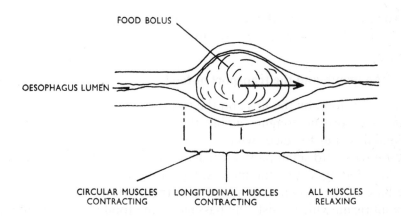

FOOD BOLUS

OESOPHAGUS LUMEN

CIRCULAR MUSCLES
CONTRACTING

LONGITUDINAL MUSCLES
CONTRACTING

ALL MUSCLES
RELAXING

Fig. 9.4 Peristaltic movement of food.

compact masses rather than as liquid or soft pulp. The tongue has to form the food fragments into compact masses for swallowing: such a food mass is known as a bolus. A bolus, lubricated by saliva, is squeezed along the oesophagus by relaxation of the muscles ahead of the bolus and contraction of the muscles around and behind it. Food is constantly moved through the system by means of waves of muscular movement along the alimentary canal: the muscular movement is known as peristalsis.

(B) The stomach: simple and compound

Once the food material reaches the stomach, it is transformed from a solid to a liquid state and its progress is arrested until this transformation is complete. The action is achieved by the churning of the stomach contents, produced by rhythmic contractions of the three muscle layers of the stomach wall and by chemical breakdown: the latter process will be considered later.

The ruminant stomach

The true stomach in cattle and sheep is known as the abomasum, and it functions as described above. This stomach is preceded by three other 'stomachs' which are really pretreatment sacs or pouches.

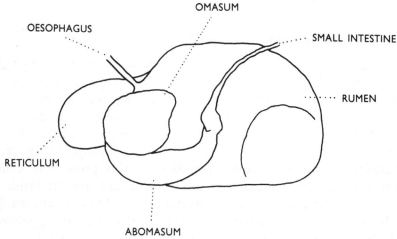

Fig. 9.5 The compound ruminant stomach.

The four stomachs together equal $\frac{1}{8}$ of the adult animal's weight, the rumen of a cow having a capacity of about 250 l. The rumen, being the largest, functions as a fermentation vat, filled with food constituents suspended in liquor. The food bolus, when swallowed, enters the rumen and either disintegrates and is dispersed in the rumen liquor or floats on the surface of the liquor and is regurgitated for further chewing (cudding). The re-chewed bolus having been pulverised into a dense mass now passes through the reticulum into the omasum. The reticulum, with horny ridges in a reticulate pattern on its inner surface, grinds down floating tough material. The liquefied and fermented contents of the rumen overflow through the reticulum into the omasum where the leaf-like projections lining the organ filter away water and allow the nutritious part of the food to pass into the abomasum for enzymic digestion.

In young suckling animals, these pre-treatment 'stomachs' have no part to play in the digestion of milk, and they are by-passed by a tube, the oesophageal groove, formed to carry liquids directly into the abomasum.

This complex stomach arrangement enables the ruminant animal to deal with large quantities of food material containing carbo-hydrates in the form of cellulose, the tough fibre in hay and straw, which without this special treatment would be indigestible and therefore of no value to the animal.

(C) The small intestine and absorption

The liquefied food material passes from the stomach through the small intestine which, in cattle, is over 40 m long and, in the pig, is nearly 20 m long. Early in its passage, digestive breakdown is complete and here the valuable food components are absorbed into the body. In order to achieve this absorption efficiently, the digested materials have to pass into and through living cells, therefore the lining consists of millions of tiny projections called villi which greatly increase the surface area and give the inside of the small intestine a velvety appearance. The special cells on the villi surfaces, which have to withstand the digestive processes whilst still remaining permeable to the products of digestion, are only short-lived. Research, in which the behaviour of these cells has been followed by radioactive labelling techniques, has shown that new cells are continually being formed at the base of each villus; each new lining cell migrates to the top of the villus in about 40 hours. The digestive tract tissues have a very rapid rate of cell replacement whereby the cell numbers are doubled in one day. About half the matter voided by an animal consists of dead cells sloughed off the digestive tract wall. Within the thickness of the intestinal lining there are many small blood vessels and lymph canals which acquire and transport the digested materials to other parts of the body.

The digestive tract continues as the large intestine but at the transition point there is an 'off-shoot', the blind gut or caecum, ending with the appendix.

(D) The large intestine

This portion of the digestive tract, also known as the colon, is larger in cross-section but shorter in length than the small intestine; in cattle, it is just over 10 m long. Some nutrient absorption may occur as the liquid food material passes along this section but the main function of the large intestine is to absorb water from the undigested residue so that by the time the last section of the tract (the rectum) is reached, this residue, known as faeces, is semisolid and in this form is voided from the animal.

(E) Rumen fermentation

The rumen contents consist of a 'soup' made up of food materials, water provided by saliva, and micro-organisms. A cow, in the course of a day, produces about 190 l of saliva and a sheep about 10 l and this makes up the rumen contents to 85–90 per cent water. In this warm, constantly mixed mass, micro-organisms thrive and in one millilitre of liquor there are likely to be about 100,000,000 bacteria and 1,000,000 protozoa. The saliva also contains bi-carbonates and phosphates which buffer the acidity of the rumen liquor to pH 5.5–7.2.

The efficiency of the fermentation process is influenced by the physical form of the food being ingested and by its balance between tough resistant material and soluble concentrated chemi-

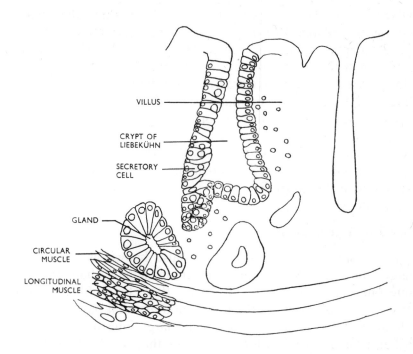

VILLUS

CRYPT OF
LIEBEKÜHN

SECRETORY
CELL

GLAND

CIRCULAR
MUSCLE

LONGITUDINAL
MUSCLE

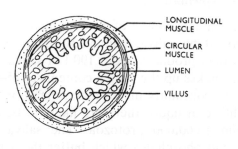

LONGITUDINAL
MUSCLE

CIRCULAR
MUSCLE

LUMEN

VILLUS

Fig. 9.6 Transverse section of small intestine (rat).

cals. Finely ground stodgy food forms a thick porridge which is difficult to churn around, makes the animal feel full and reduces appetite. At least 30 per cent of the food fed needs to be fibrous to enable the rumen to work. The micro-organisms are influenced by the chemicals contained in the food and are less efficient if more than three-quarters of the food is made up of concentrated materials.

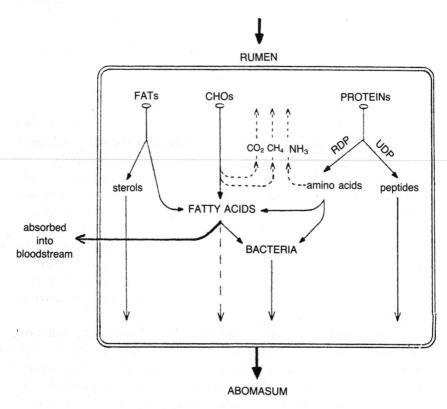

Fig. 9.7 Reactions in the rumen.

About 70 per cent of the digestible dry matter of the food is broken down by micro-organisms. They break down the cellulose walls of plant cells to release the cell contents and convert the cellulose to fatty acids, a large proportion of which are absorbed into the bloodstream through the rumen wall.

Tough complex proteins derived from animal sources, heat-treated products or stored food where formalin has been used, are described as being undegradable (UDP) and pass through the rumen largely unchanged, but proteins in grass and plant-derived foods are easily broken down and degraded (RDP) into their constituents of organic acids and ammonia. Active micro-organisms, well supplied with energy from other food sources, use the degradation products to multiply, thereby fixing the protein derivatives into a form which, in the abomasum, will be digested as protein.

(F) Chemical digestion: enzyme action

All the chemical breakdown reactions in digestion are carried out by enzymes. An enzyme is a type of catalyst which brings about a chemical reaction if certain conditions prevail. Any one enzyme acts specifically on one material and only in particular conditions.

To enable an enzyme to act, there must be a ready supply of the compound to be broken down, conditions of temperature and acidity must be satisfactory, and the products of the enzyme-assisted reaction must be dispersed. The reaction is reversible in that its progress is controlled by the relative amounts, or concentrations, of the intact chemical compound and the break-down products. Most chemical processes in the animal body are carried out in this way. The enzymes which act in the digestive system are many in number, and only the main ones will be dealt with here.

In the mouth the saliva, which has a lubricating function, contains the enzyme ptyalin; when ptyalin is mixed with the food by the action of tongue and teeth it commences the breakdown of starch to sugar. The ptyalin continues to act in the stomach. The presence of food in the stomach stimulates the wall to secrete gastric juice which contains strong hydrochloric acid to dissolve fibre and hydrolyse fat, pepsin, which starts the breakdown of complex proteins, and rennin, which coagulates milk.

In the ruminant animal, the bacteria in the rumen break down cellulose and starch and release fatty acids; other micro-organisms, such as protozoa, thrive and multiply on these products to form animal protein which is absorbed later in the digestive process. The ruminant salivates continuously and the saliva contains urea, extracted from the blood, which as a raw material of protein and amino-acids supplements the micro-organisms' diet. The animal therefore derives essential amino-acids from the synthesis carried out by rumen organisms.

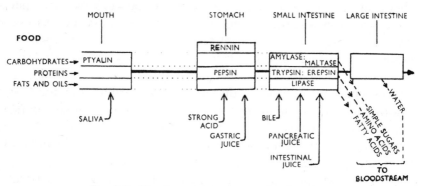

Fig. 9.8 The important enzymes in digestion.

The food material passes next into the small intestine where, at the outset, it is mixed with bile. Bile is produced by the gall bladder in the liver and contains greenish iron compounds from the breakdown of haemoglobin, and is very alkaline. The change from very acid to very alkaline conditions terminates the activity of the enzymes so far encountered, and produces conditions in which intestinal enzymes can act.

The pancreas, a secretory gland situated near the first section, pours pancreatic juice into the intestine through the pancreatic duct. This contains amylase which continues carbohydrate break-down, trypsin which breaks down proteins by a further stage and lipase which splits fats and oils emulsified by the alkaline bile, into fatty acids and glycerol. The chemical breakdown is finished by enzymes secreted by glands in the intestinal lining at the base of the villi. Invertase acts on complex sugars to produce very simple one-molecule carbohydrates which can pass through the lining cells. Erepsin reduces the protein materials to amino-acids which can be absorbed, and lipase converts any remaining fat to fatty acid and glycerol—products which can be absorbed.

As the food material passes along the small intestine, digestive processes are completed; the simple products pass into the body system and are absorbed. These energy-containing chemicals and raw materials for new cells now have to be transported by the bloodstream to the point where they are needed.

For economic animal production, not only must the right foods be fed but the right amounts must also be fed. By knowing the chemical analysis of a food and its approximate digestibility, that is the proportion that will be available to the animal in the process of digestion, it is possible to estimate the feeding value of a food numerically. The energy value of a food is stated as its metabolisable energy, measured in joules per kg of food. The energy content of food is discovered by burning the food in a calorimeter and comparing the results with those obtained from animal feeding trials. The protein value of a food is stated as the percentage of the food which is digestible crude protein found from analysis and digestibility trials.

By knowing the daily requirements of an animal of particular type or age, from figures of metabolisable energy and digestible crude protein, it is possible to compose a ration to provide these constituents in correct amounts thereby avoiding waste. It must however be remembered that an animal can only deal with a total quantity of food per day having a dry matter total equal to about 3 per cent of the animal's body weight. On the other hand, the day's ration cannot be concentrated into a convenient pill because the digestion becomes inefficient if insufficient bulk of material is fed.

10 Blood and the circulatory system

The circulation of the animal is a transport system whereby the raw materials and oxygen, which all living cells need, are supplied to every cell in the body. Harmful waste products from living processes are safely removed and dealt with by this system.

(A) The constituents of the blood and their functions

Firstly, consider this mixture of chemical substances, living cells, and water—the blood. When blood is examined under the microscope, the red disc-like blood cells are the most obvious features. On careful examination, occasional irregular and colourless cells will be seen: these are the white blood cells. All the living cells are floating in a liquid which, when separated from the cells, is pale straw-coloured and watery—the plasma or lymph.

Red blood cells

These cells are all the same size, 0.01 mm in diameter; the same shape, circular and dished on each side, and, unlike most living cells, are without nuclei. Without nuclei, these cells cannot reproduce: new cells are produced by the marrow tissues of the bones. Each red blood cell has a functional life of twelve to sixteen weeks, after which it is broken down in the liver. The most important feature of these red cells is that they contain haemoglobin which is a red iron-containing compound. In the presence of oxygen, it forms a loosely combined compound, oxy-haemoglobin, by which oxygen is carried from the lungs to the tissues of the body and released to the cells. Since haemoglobin contains iron, shortage of iron results in shortage of haemoglobin and the animal is then anaemic.

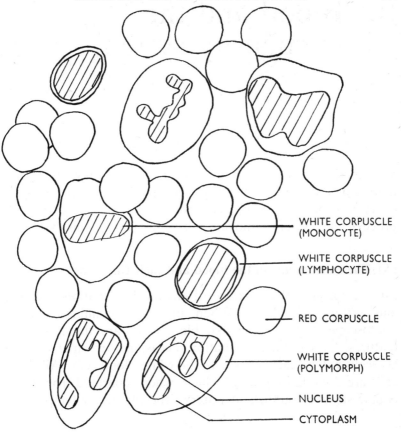

WHITE CORPUSCLE
(MONOCYTE)

WHITE CORPUSCLE
(LYMPHOCYTE)

RED CORPUSCLE

WHITE CORPUSCLE
(POLYMORPH)

NUCLEUS

CYTOPLASM

Fig. 10.1 Types of blood cell (human).

White blood cells

There are four hundred times as many red blood cells as there are
white but, in spite of the fewer number, the white cells have an
important role to play in controlling disease. The white cells are
formed in the lymph glands distributed around the body and are
flexible cells containing nuclei. When the animal body is invaded
by disease-causing organisms at any point, the white blood cells
move to the infected area and can pass through the walls of the
blood vessels in order to overcome the invaders.

The plasma

In the blood, plasma makes up 55 per cent of the volume. It is mainly water, but contains some very important constituents. One of the important chemicals in solution is sodium carbonate which combines with carbon dioxide produced by living cells; the carbon dioxide is thus carried to the lungs for excretion.

The plasma carries the product of digestion to the cells of the body and returns with urea, the nitrogenous waste product of worn-out cells.

In addition to the nervous control system of the animal body, there is a chemical control system and the chemical messengers, or hormones, also circulate with the plasma of the blood.

The plasma also contains blood proteins which, as will be seen later, have an important function in the clotting mechanism which seals broken blood vessels.

Although the circulation of the blood has to be rapid and efficient in order to meet the body cell demands, it could also allow the rapid spread of any disease organisms which might gain entry. An efficient precaution against this occurrence is provided by the presence, in the blood, of special chemicals known as agglutinins and antibodies. These chemicals cause any 'foreign' organisms (or proteins) to be clumped together and destroyed.

The functions of the blood may therefore be summarised as follows:

1. Transport of cell requirements and by-products:
    Oxygen from the lungs to the tissues.
    Carbon dioxide from the tissues to the lungs.
    Nutrients to the tissues from the digestive tract and liver.
    Nitrogenous waste from the tissues to the kidneys.
2. Chemical co-ordination by hormone circulation.
3. Disease control.
4. Temperature regulation. Circulating blood carries heat which can be dissipated at the surface, if necessary.

(B) Blood vessels and the heart

Movement of blood

The cells of the body are so numerous that they are not individually supplied with blood; in the tissues, however, there are many small blood vessels—so minute that the blood cells pass along singly—which bring blood very close to the cells. The final link in the supply of nutrients and removal of waste products is provided by the lymph or plasma which can diffuse through the walls of the blood vessels and bathe every cell. These tiny blood vessels are known as capillaries and they form a network through every organ and tissue. The flow along these capillaries is controlled by muscle 'valves' at each junction, and can be increased, decreased, or diverted to a particular part.

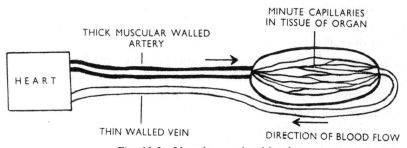

MINUTE CAPILLARIES IN TISSUE OF ORGAN

THICK MUSCULAR WALLED ARTERY

HEART

THIN WALLED VEIN

DIRECTION OF BLOOD FLOW

Fig. 10.2   Vessels carrying blood.

Considerable pressure is required to move blood through a fine capillary network; this pressure is provided by the heart. When this blood is collected into large vessels again, the pressure is very low because it has been almost exhausted in the process of forcing the blood through the capillary network. Usually, therefore, the blood only passes through one set of capillaries and then returns to the heart. The large blood vessels carrying blood from the heart to the capillaries, and back to the heart again, usually lie alongside each other in the body: these vessels are the arteries and veins.

Arteries

Arteries are the tough blood vessels which carry the pressurised blood from the heart to capillaries in all parts of the body; they are thick-walled, muscular, and elastic. The heart, with its rhythmic pulses, produces a fluctuating pressure surging through the system. If this peak pressure were applied to the fine capillaries they would rupture, and minute internal bleedings would occur; these haemorrhages can be serious in vital organs such as the brain. The arteries, therefore, expand in response to the pressure peak and, by off-beat contraction of the muscular artery wall, they even out the pressure surges; a constant blood pressure is thereby produced which forces blood through the capillaries.

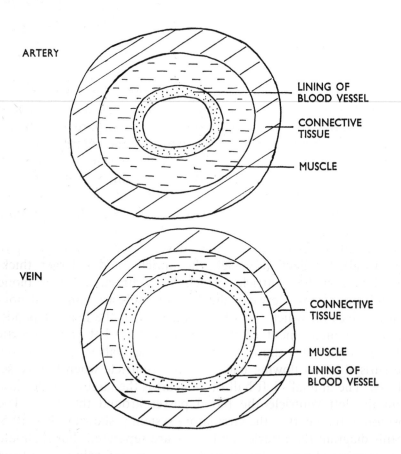

Fig. 10.3 Transverse section of artery and vein.

Veins

The fine network of capillaries gradually joins up to form large vessels collecting blood from an organ or tissue, and in these vessels—the veins—blood completes its circulation back to the heart. The blood in the veins is at very low pressure which does not fluctuate, so the vein walls are thin and inelastic. The pressure in the veins is often so low that there is a tendency for the blood flow to reverse direction: this is prevented by valves. Valves, which consist of three pockets around the inside wall of a vein, are situated at intervals along the main veins. Any back-flowing blood fills the three pockets and closes the vein, normal flow flattens the pockets against the wall of the vein, and blood passes freely.

The heart

Although most essential organs in an animal are duplicated to safeguard against disease or injury, the animal has only one heart; yet, considering the job it has to do, it may be regarded as one of the most important organs of the body. From before birth and throughout the life of the animal, it pumps constantly—rebuilding and replacing its muscle while working, adjusting its pulse rate and size to the body's varying demand for blood, and all the time pumping a liquid in which are floating blood cells which are more fragile than eggs in water.

The heart consists of two similar halves. Each half is made up of a thin-walled collection vessel, the auricle, and a larger thick-walled vessel, the ventricle, which by contraction forces blood around the system. Between these two vessels are important non-return valves on which the whole efficiency of the heart depends. At the commencement of the arteries are pocket valves (aortic valves) similar to those in veins; these valves prevent the pressurised blood flowing back into the ventricle when this vessel relaxes. The animal's left half of its heart is the larger side because from the left ventricle the blood passes around the body. The position of the heart in the circulatory system is shown in Fig. 10.5. In this diagram the arteries and veins are separated, but it should be remembered that in the animal the artery and vein to and from an organ lie close together and are usually situated near bone (especially the backbone) for protection.

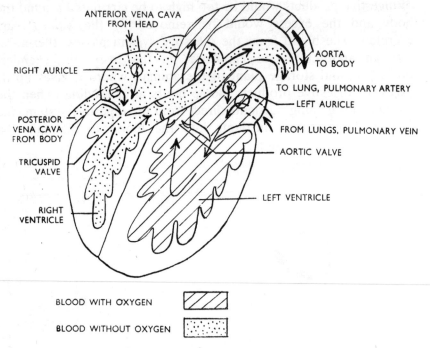

Fig. 10.4 Diagram of heart.

(C) The circulation

In one complete circuit of the animal body, the blood passes through the heart twice. From the left side of the heart the blood travels through capillaries in one of the body organs and then returns to the right side of the heart. The second part of the circulation carries the blood through the lung capillaries where it gives up carbon dioxide and absorbs oxygen; the blood then returns to the left side of the heart to repeat the process. All the blood is oxygenated once every circuit. Around the body system, the blood may follow any one of many routes. A portion of the circulating blood passes through the kidneys where unwanted impurities and excesses are filtered out and excreted as urine. A large proportion of the blood is routed via the digestive system, and on this route a variation on the simple system is encountered— a portal system. Here the blood first passes through the intestinal capillaries and acquires digestive products. After feeding, the level of digestive products would be too high to be circulated around the body and the excess would be removed by the kidneys and excreted. The blood from the digestive system passes, therefore, through another set of capillaries in the liver, the chemical processing and storage plant of the animal, where it gives up its excess of digestive products. Some hours after feeding, when the blood is not gaining nutrients direct from the digestive system, the liver releases stored nutrients into the blood to maintain the constituents of the latter.

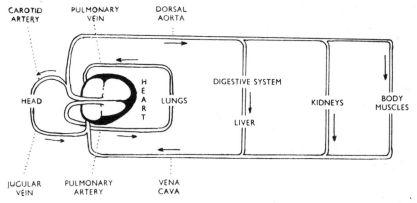

Fig. 10.5 Diagram of blood circulation.

The blood rapidly travelling around the animal body keeps the millions of living cells in an environment which is ideal for efficient activity.

(D) The mechanism of blood clotting

From the foregoing study of the blood, it is obviously extremely important to the animal that loss of blood should not occur.

If, by severe injury, an artery is ruptured then some quick emergency action is called for on the part of the stockman, otherwise the heart will pump away all the blood. The most effective form of first aid is to apply pressure to the artery, on the heart side of the injury, and so stop the flow of blood. The veterinary surgeon's skill will be necessary to repair the torn artery and close the wound.

There are many occasions when bleeding is caused by only

superficial injuries, and under these conditions the blood system is 'self-sealing'. The blood contains a protein named fibrinogen, an enzyme which causes fibrinogen to solidify, and an inactivator to this enzyme.

When the wall of a blood vessel is damaged, a neutraliser for the inactivator is produced and this allows the fibrinogen to be solidified by the clotting enzyme. Fibrinogen solidifies to form a network of fibres amongst which the minute platelets in the blood form a solid mass and prevent further loss of blood.

Since this process is caused by secretions from the damaged blood vessel wall, clotting occurs more readily in a lacerated wound than in a clean-cut injury.

11   Animal reproduction

The farmer's success is very dependent upon the breeding potential of his livestock and on the ability of their offspring to thrive and grow. In order to overcome the problems which arise, and to produce ideal conditions, he must understand the biological processes involved. Farm livestock may be studied as a group since all are mammals and are basically similar.

(A) Male reproductive organs

Each of the male gametes, or sex cells, has a head which is nearly all nucleus and a tail three to four times the length of the head: these cells are known as spermatozoa. They are formed by the

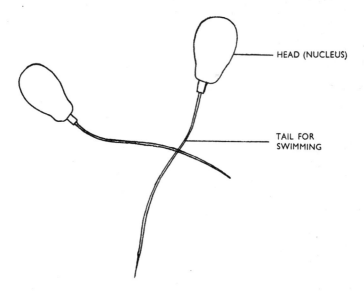

HEAD (NUCLEUS)

TAIL FOR SWIMMING

Fig. 11.1   Bull spermatozoa.

lining cells of the coiled tubes which form the testis. In their formation by the process of meiosis, or reduction division, the male spermatozoa each contain half the chromosomes necessary for the production of a new animal. By eventual fusion with the female sex cell, or ovum, these male chromosomes are matched with a similar number of chromosomes produced by the female in the ovum; a complete set of chromosome pairs is thereby formed and this contains all the genetic make-up of a new animal. At the base of the testis is the epididymis in which the continuously produced spermatozoa are stored. Because of their importance for continuance of the species, these structures are duplicated. Sperm production requires a temperature below that of the rest of the body, and therefore the testes are suspended externally in the scrotum. The testes are attached to the spermatic cord which

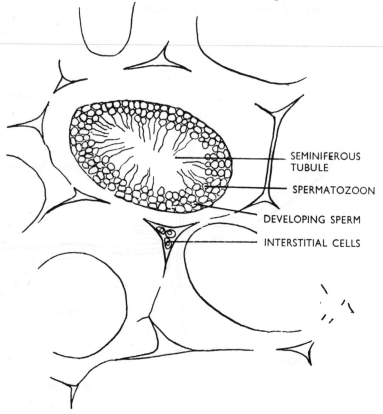

Fig. 11.2 T.S. testis (rat).

contains blood vessels and the vas deferens, the tube leading from the epididymis into the abdominal cavity. When mating takes place and ejaculation occurs, spermatozoa pass along the vas deferens and are mixed with secretions from the accessory glands to form seminal fluid in which the sperms are activated and nourished. The vas deferens join with the urethra, the urine tube from the bladder to the tip of the penis; by means of the rigid penis inserted into the female, the semen passes down the urethra and is deposited at the inner end of the vagina. The semen contains many million sperms per cubic centimetre, but only one sperm can fertilise one ovum. The semen production of farm animals varies and the natural mating process seems very wasteful; dilution and preservation techniques, which are continually being improved, are therefore used and enable many artificial inseminations to be carried out with the product of one natural ejaculation.

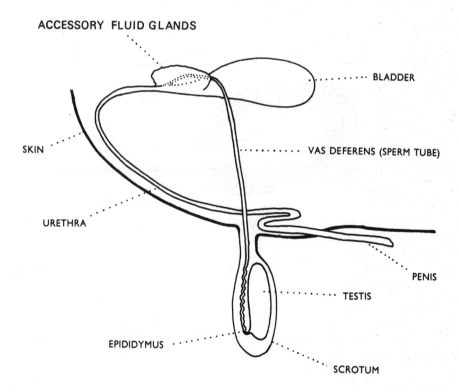

Fig. 11.3 Male reproductive organs.

PRODUCTION OF SEMEN BY FARM ANIMALS

	Accessory Gland Secretions	Total Volume
Horse	90 ml	100 ml
Cattle	4 ml	5 ml
Sheep	less than 1 ml	1 ml
Pig	200 ml	250 ml

(B) Female reproductive organs

The male reproductive system is merely comprised of organs devoted to producing seminal fluid containing living spermatozoa. This is simplicity itself compared with the female system which not only produces ripe egg cells, or ova, with rhythmic regularity but also protects and nourishes the young animal in a very special environment before birth and feeds it after birth. Later in this section, the control of these interacting activities will be studied, but first consider the system of organs in the female animal.

The pair of ovaries are attached to the roof of the abdominal cavity and to the rear of the kidneys; one ovary being situated on each side of the mid-line. The ovaries appear as small lumpy glands; under hormone stimulation, once the animal is sexually mature, egg cells (or ova) develop in the substance of the ovaries. Each ovum is surrounded by a ball of cells called a Graffian follicle. As each follicle matures and increases in size, it migrates to the surface of the ovary and, when ripe, ruptures to release the ovum and some watery albumen. Close to the ovary is the funnel-shaped opening to the Fallopian tube, or oviduct, into which the ovum and secretions pass.

As the Graffian follicle matures and rises to the surface of the ovary, it produces hormones which initiate and maintain the signs of heat, the female animal's attraction of the male, and acceptance of mating. After rupturing, the cells which formed the follicle become the corpus luteum, or yellow body, which produces another hormone; this hormone arrests the development of further follicles and conditions the uterus wall to receive the fertilised ovum.

In the upper oviduct, the albumen coagulates around the ovum and forms a plug. When an animal on heat is mated, semen is deposited at the inner end of the vagina by the male. If artificially inseminated, much-diluted semen is injected into the uterus by the inseminator who feels through the rectum wall and guides the insemination tube through the cervix—the narrow neck of the uterus. The sperms, activated by the accessory gland secretions and sustained by uterine secretions, are chemically attracted to swim along the length of the uterus, through the oviduct to the ovum cell, and to accumulate in the albumen plug.

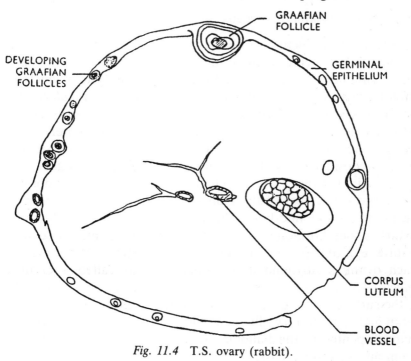

Fig. 11.4 T.S. ovary (rabbit).

Sperms in the vagina survive only for about an hour, but those travelling through the uterus and oviduct may take up to twenty-four hours to reach the ovum. This is a considerable journey—in the cow it covers a distance of 375–450 mm—for microscopic sperms, and many die on the way.

After a few hours the albumen plug around the ovum dissolves and the ovum, after one sperm head has entered and fused with the ovum nucleus, becomes impenetrable to other sperms. The

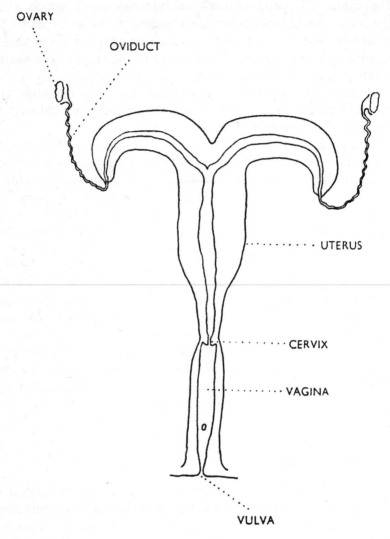

OVARY

OVIDUCT

UTERUS

CERVIX

VAGINA

VULVA

Fig. 11.5 Female reproductive organs (from above).

fertilised ovum now travels slowly down the oviduct, taking several days to reach the uterus; by this time it has become a ball of cells produced from the original fertilised ovum by repeated cell divisions. In the uterus this ball of cells becomes embedded somewhere in the lining of the uterine wall.

The uterus is a thick-walled muscular container which is T-shaped; the thin oviduct tubes are attached to the arms of the 'T' and, at the base, the cervix—a muscularly constricted opening—protrudes into the inner end of the space of the vagina. The vagina is a tube passing through the pelvic bones to the outer opening, the vulva, situated under the tail and below the rectal opening. The cervix and vagina have to be capable of considerable dilation to permit the young animal to be born.

In an animal such as a cow or ewe, where only one or two young develop at one time, the ova are produced from each ovary alternately. If the normal is to produce single offspring, twinning is an inherited character. Single young are developed and carried in the main trunk of the uterus. In animals which produce large numbers of young at one time, for example sows and rabbits, both ovaries release ova when heat occurs and the young develop in both arms of the uterus. More ova may be fertilised than the female can carry—up to twenty in a sow—and subsequently the unwanted fertilised ova are reabsorbed. The numbers of young which develop depend on the condition of the animal, and how many it can nourish and carry in the uterus.

When identical pairs of young are produced—strictly speaking mirror images of each other—one fertilised ovum has split in two. If the split is incomplete, Siamese twins are produced—some structure or tissue being common to both.

(C) Pregnancy and birth

Once a developing animal, or embryo, has become implanted in the uterine wall, membranes develop to enclose it. As it increases in size, the embryo becomes free to move in the uterine cavity but is cushioned for protection by liquid present between the layers of the three membranes within which it is enclosed. Since all the raw materials which are necessary for the development of the young animal must be derived from the mother, a special organ develops temporarily; this organ, known as the placenta, transfers these essential materials from the mother to the embryo.

After about one third of the gestation period has elapsed, the heart and rudimentary blood system have developed in the embryo and major blood vessels arise from the centre of the abdomen and carry blood to and from a set of capillaries developed on the placenta. These capillaries come in close contact with a matching set of capillaries containing the mother's blood and situated on the uterine wall. No intermixing of the mother's blood with that of the embryo occurs, but substances are easily passed in both directions and this link continues until birth takes place. Once the placenta is developed, it produces a hormone which is responsible for continuing the pregnancy; it is the withdrawal of this hormone which brings about contraction of the uterine muscles and the birth of the young.

The onset of birth is indicated by recognisable signs in the

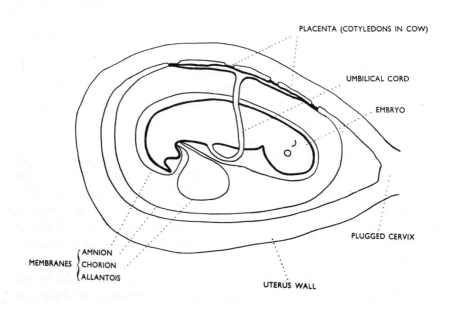

Fig. 11.6 Young animal in uterus.

animal. The mother or, more appropriately, the dam becomes restless and seeks seclusion; milk will usually be present in the udder in large quantities; and, immediately before giving birth, slackening of the ligaments around the pelvic bones prepares for the passage of the young through the pelvic cavity. The actual birth of the young is assisted by the presentation first of the narrowest parts of the body—usually the muzzle and fore-feet; this provides a wedge to dilate the passage through which the young must come. The membranes surrounding the young animal rupture early in the process, and the liquids which they contained act thereafter as lubricants. The rhythmic contractions of the powerful uterine muscles force the young through the continuous tube formed by the cervix and the vagina. Soon after birth, continued contractions expel the placenta and any remaining membranes.

At birth the oxygen supply which the foetus has derived from the blood circulation through the placenta is cut off; the newly-born young animal must, therefore, immediately start to breathe. While developing in the uterus, the young animal's blood circulation to the inoperative lungs has been by-passed; at the moment of birth, this by-pass closes and the full circulation is established.

Although the young will suckle the dam's milk for some weeks, for the first day of life the offspring is able to absorb into its body system antibodies conferring disease resistance which are contained in this milk; during this period the dam's milk is particularly rich in antibodies and other essentials.

When born, the young animal has a flexible skeleton and is only thinly fleshed with complete absence of fat. This latter feature causes the young animal to be sensitive to cold and, therefore, conditions which are ideal for birth include both hygienic cleanliness and warmth. The least disturbance of the dam whilst in the process of giving birth, is also desirable unless abnormalities occur.

The whole of these reproductive activities is controlled in the dam by hormones, and the balance of the hormones produced by the various organs is critical. Fig. 11.7 shows the various organs and hormones involved in the progress of a normal pregnancy.

If mating and fertilisation of the ovum do not occur, the corpus luteum degenerates and further follicles develop in the ovaries.

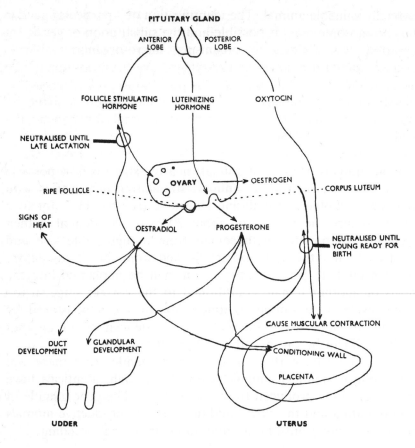

Fig. 11.7   Female hormone functions.

(D) Genetic manipulation and embryo transfer

The development of the laboratory techniques which enable ova, sperm and embryos to be viably maintained outside the animal body and which allow fertilised embryos to be manipulated under the microscope, stored indefinitely by freezing and re-introduced into the uterus, have led to far-reaching possibilities.

Working with the newly fertilised embryo it is possible to remove the pro-nucleus provided by the sperm and replace it with a nucleus from the dam so that the embryo now develops with two identical sets of genes so eliminating genetic variation of that mating and concentrating the genetic characteristics of a commercially valuable animal. The identification of a particular gene in the chromosome mass is possible and the substitution of genes has imparted new abilities to organisms, micro-organisms able to produce animal hormones and enzymes have been developed. To the early embryonic ball of cells it is possible to graft in groups of cells from other animals so a half goat/half sheep is living an apparently normal life; a 'tailor-made' animal for commercial production is a possibility.

The production of large numbers of embryos from highly valuable animals or even of animals nearly extinct is now possible by selecting the high value animal, stimulating her ovaries with injections of Follicle Stimulating Hormone (see fig. 11.7) for four days to secure the release of about 20 ova. The animal is then artificially inseminated with semen from a high value sire and seven days later the embryos are flushed from the uterus before they can implant. These embryos can then be implanted into the uterus of another female conditioned to the correct stage in the oestral cycle by hormone treatment, or they can be stored by freezing for later use. The receptor female can be of another species and can be on the other side of the world.

Micro-surgically, early embryos can be divided and these will then produce identical animals (clones) and eight identicals have been successfully produced from one embryo. The gene gamble of natural mating and the slow build-up of highly productive animals are fast becoming out-of-date limitations in animal breeding.

12 Growth and maturity of farm animals

A question which is often asked is, 'How can one tell the age of an animal by simply looking at it?' When the age of an animal is estimated by anyone with experience, two main features are used as indicators. Firstly the animal's size and, secondly, the relative size proportions of various parts of the body; a young animal appears to have a large head and long legs, while age brings thickening of the body.

These two features indicate that two changes occur simultaneously as an animal ages. In the matter of size, often measured as liveweight, the increase is known as growth; this is very rapid in the young animal, but slows steadily with advancing age until the maximum size for the type and breed is reached.

The second feature, the variable growth rate of different parts of the body, is known as development. The degree of development of any part is influenced by both the genetic potential and the level of nutrition at the time that particular part is being developed. An animal which is poorly fed after weaning often retains young animal proportions, because head and limb development are normal but body development is stunted. Development occurs as waves of growth; these start with the extremities and reach the loin region at maturity.

The raw materials derived from the food are used by the animal body for the development of different tissues and these are classified in a definite order of preference as shown in fig. 12.1.

The amount of raw materials used for each purpose will depend on the growth speed of that tissue. While the brain and bones are developing rapidly and utilising most of the available raw materials, there is unlikely to be much raw material free to produce muscle and fat. Later in life, by adjusting the amounts of food fed to the animal, the relative development of parts of the body can be influenced. A young over-fed animal will develop muscle well, but if food quantities are then restricted little fat will be formed. If, on the other hand, food is restricted in early life and later is generously increased, early developed muscle will be poor but the animal will deposit large quantities of fat.

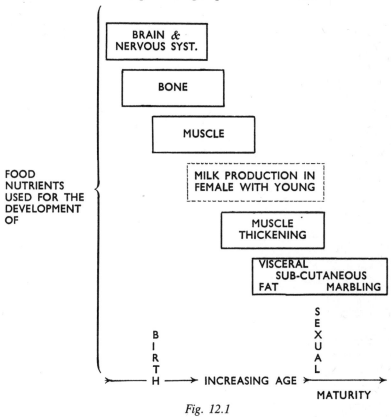

Fig. 12.1

Since, in a meat animal, the proportion of fat to muscle very much affects the quality of the meat, a closer study of these features is worthwhile. The ideal meat quality is found where the muscle consists of fine fibres, and is dark in colour thereby indicating good flavour; small quantities of fat are dispersed among the muscle fibres to assist the cooking quality of the meat.

Muscle development

Most of the muscle tissue is formed by the time the animal reaches the age of sexual maturity; subsequent muscle growth consists of an increase in the thickness of the muscle fibres and the meat becomes coarser. Exercise increases this tendency. When the muscle is worked, an increased blood supply causes more haemoglobin to be present in the muscle and this results in a darker colour and more flavour. The maturing of the muscle is delayed by castrating male animals, by a lack of exercise, and by a deficiency of iron.

Fat development

From approximately the age of attaining sexual maturity until the animal is full-size, a decrease in the development of other body parts results in the deposition of fat. Fat is an insoluble carbohyrate-like compound deposited as an energy reserve which may be mobilised and used to produce energy at any time. Three stages of fat deposition occur as an animal matures. In the first, fat is deposited as thick masses around internal organs; this visceral fat is of little value in respect of meat quality. The second stage occurs when fat is deposited in a soft layer beneath the animal's skin. This subcutaneous fat gives the animal a smooth rounded appearance but, when slaughtered, gives fatty joints of meat. The final stage is the development of fat within the muscle, between the muscle fibres, producing 'marbling' in the meat. This final stage is very dependent on the skill of the farmer. The subcutaneous fat stage can easily be seen and felt, but only the skilled man knows how much longer to feed the animal and can judge the right moment for slaughter so as to produce the ideal meat quality.

Carotene, a yellow pigment, is deposited in fat and the yellow colour sometimes detracts from the quality of the meat. The Channel Island breeds of cattle have a high carotene yellow fat; it may also be induced in other cattle if fattening has been inconsistent and at some stage deposited fat has been utilised for energy—the carotene remaining and accumulating in the meat.

Although the foregoing maturation process is a fairly simple and straightforward one, there is no other animal production process which is so influenced by conditions. Success in the production of the ideal product is, therefore, very dependent on the skill of the farmer.

Section IV

The environment

Introduction

The environment is the place in which an animal or plant must grow.

In the wild state, plants are found in certain habitats where the environment is favourable to their growth. If plants are to be grown successfully by a farmer, then he must simulate, or even improve upon the conditions to be found around a plant in the wild state.

Wild animals will be found living in places where their natural food and shelter can be provided. Domesticated animals require protection and a carefully selected diet if they are to thrive.

In their natural conditions plants and animals are found in small colonies scattered throughout the countryside, but plants on farms are gathered together in small spaces. As many as 75,000 sugar beet plants are grown on a hectare of soil. Under conditions like this, the natural pests and diseases of a plant or animal build up rapidly.

Successful animal and plant production depends upon controlling the environment within the limits of tolerance for satisfactory growth.

A study of the environment helps us to know better what we can do to improve output. It also helps us to appreciate that there are many things in the air and soil which we cannot alter, and must therefore accept as natural limitations.

During our struggle to improve output from the land, we must remember that the many different soil types, variable water content and the effects of altitude, combine to produce a multiplicity of specialist environments which encourage the growth and development of wild flowers, butterflies and birds which make the countryside so beautiful and enjoyable for all.

13 Meteorology

What is meteorology?

Meteorology is the study of the weather and factors which cause changes.

The nature of the surface

The surface of the earth is covered by air, water, and soil. These are held to the earth by gravitational force but are moved about by the earth's rotation, the gravity of the moon, and the heat from the sun.

The movement of air

Air becomes heated by the sun and rises by convection. Hot air can hold more water vapour than cold air, so this rising air can

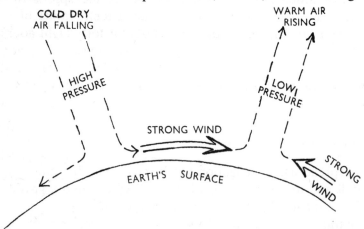

Fig. 13.1   Movement of air.

pick up moisture. The air movement creates winds, so the moist air will move. As the air cools again it can no longer hold its additional moisture, so rain falls. This process occurs continually on the surface of the earth, and is the source of water for plant growth.

The importance of meteorology to the farmer

Rain makes plants grow and provides clean drinking water for man and animals, but many farming operations are best carried out when the weather is dry. For example, ploughing, seed bed preparation, and seed drilling require dry weather especially on clay soils. Hay-making and harvesting in warm dry weather cuts down the need for artificial drying which can be expensive. Any help which the farmer can receive from instruments or cloud formations—which might indicate changes in the weather—must be accepted and appreciated.

Farmers are usually reasonably good weather prophets. This ability stems from long experience of local signs, behaviour of animals, birds and insects, and the movement of cloud.

Blight forecasting

In more recent years local meteorological recording stations have forecast the conditions likely to bring about the spread of the fungus disease, 'blight', on potatoes. When a temperature of 10°C and a relative humidity of 75 per cent exists for forty-eight hours, a Beaumont period is declared and the prevalence of blight is increased for the following three weeks. Some farmers wait for Beaumont periods before carrying out their preventative spraying programme.

Irrigation need

The widespread use of irrigation has brought with it a close watch on rainfall and terms like 'irrigation need' which help a grower to plan the use of his equipment to the best effect.

The nation's meteorological service is the subject of much criticism and little praise. We forget the times when the forecast is correct, but choose to remember when the weather was not exactly what was expected. In the service there is a special branch for helping farmers. The men who operate this are experienced in the specific problems which farmers are likely to meet. A telephone call to their office would often save a lot of time.

(A) Climatic factors and recording

To obtain the complete picture of the weather conditions at any time or during a growing season for example, several instruments must be consulted and recordings taken. The following records might be required: pressure, rainfall, evaporation, humidity, maximum air temperature, minimum air temperature, soil temperature, wind direction and speed, sunshine duration, and cloud cover.

A brief study of each of these in turn will reveal their value to the farmer.

Pressure

The air is a mixture of nitrogen gas, oxygen gas, and water vapour, with small amounts of other gases mixed in. All the constituents are in gaseous form and their individual chemical particles, called molecules, are extremely active. The molecules are constantly on the move, bumping into one another, bombarding the objects on the earth, penetrating every crevice in the soil, and even attempting to fly off into space. Just as a stone thrown into the air will fall again, so a molecule moving upwards is eventually slowed down and drawn back to earth by gravitational force. Even so, some air has been detected as high as 135 km up, but the majority of the air or atmosphere is concentrated above the earth in the first 7–15 kilometres.

The result of this constant molecular bombardment is called pressure. Since there are most molecules near the earth, that is where pressure is greatest and at high altitudes pressure falls.

Air pressure is powerful; it can lift water as in a syphon or a lift pump.

Air presses equally in all directions. It cannot move downwards because the earth becomes too dense; it cannot escape upwards

because gravity pulls it back, so it presses sideways around the
world. Should there be a drop in pressure anywhere, air will rush
in to fill up the space.

The pressure of the air remains fairly constant but even small
fluctuations affect the evaporation of water. Reducing the pressure
of air above water helps the water to evaporate. Increasing the
pressure of the air restricts the evaporation of water. Sensitive
instruments register pressure changes however slight they may be.
The human eye is sensitive to air pressure: one's eyes seem to
protrude further when the pressure is low. The human ear is
sensitive to pressure: a sudden change of altitude can cause pain in
the eardrum if the pressure on either side of the drum becomes
unequal.

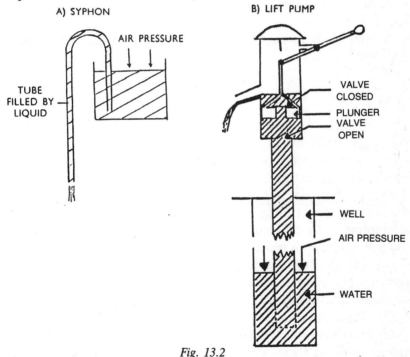

Fig. 13.2

Pressure measurement

To measure pressure accurately a barometer is used. Air pressure
will support a column of water about 10 m high, and a column of
mercury 760 mm high. A simple barometer can be made by filling

a thick glass tube, about 1 m long and closed at one end, with mercury liquid. Invert the glass tube and place the open end in a dish of mercury, without allowing mercury to escape, until the open end of the tube has been submerged. The mercury will drop a little when released as the column exceeds 760 mm. The space at the top of the tube is a vacuum.

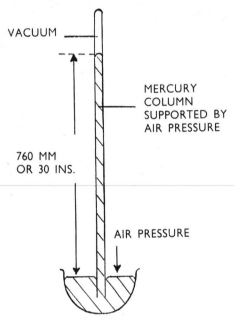

VACUUM

MERCURY
COLUMN
SUPPORTED BY
AIR PRESSURE

760 MM
OR 30 INS.

AIR PRESSURE

Fig. 13.3 Mercury barometer.

As the pressure varies from day to day the height of the mercury column varies and can be measured with a ruler and recorded in millimetres of mercury; 760 mm of mercury, or over, is a high pressure, and 750 mm of mercury and below, is a low pressure.

The value of the barometer and the pressure record becomes obvious when the separate records for wind speed, rainfall, frost, and sunshine are compared with the pressure record.

The aneroid barometer is a partially evacuated cylindrical metal container which is compressed by air pressure. A lever, which indicates the amount of compression, points on a dial to high or low pressure.

The chief cause of pressure change is the heating of air by the sun, or by the back radiation of stored-up heat from the sun by a large land mass. This heated air is forced to rise by convection leaving behind it a reduced pressure.

Areas of low pressure are called depressions or cyclones if the pressure is extremely low. Areas of high pressure are called anti-cyclones. As a guide, the weather is settled and fine in an anti-cyclone but often cold in winter and spring; in a depression there is a strong wind and rain is likely.

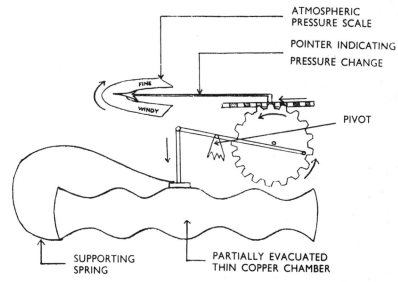

ATMOSPHERIC
PRESSURE SCALE

POINTER INDICATING
PRESSURE CHANGE

FINE

WINDY

PIVOT

SUPPORTING
SPRING

PARTIALLY EVACUATED
THIN COPPER CHAMBER

Arrows on the diagram indicate the direction of movements when pressure increases.

Fig. 13.4 Aneroid barometer.

Wind

Air movement is the direct result of pressure change. Air will rush into a depression causing strong winds to develop. In a very deep depression called a cyclone, the winds may exceed gale force and reach typhoon speeds of 150 km per hour or more.

In low pressure areas the weather is bound to be unsettled, but usually mild in the winter.

In anti-cyclones where pressure is high the winds are gentle, moving away from the area as cool air falls from above. The weather is settled, but usually cold in winter although often sunny by day. During the summer, high pressure brings warm sunny days.

Measurement of wind

Wind speed or pressure can be measured by a moving flap with a pointer attached indicating a reading on a calibrated scale. Or the speed of rotating cups can be measured by speedometer. This type of instrument is called an anemometer.

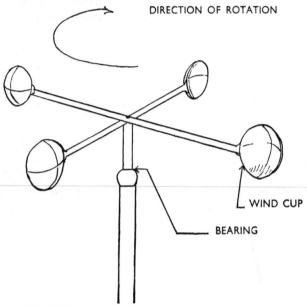

Fig. 13.5   Anemometer, windspeed measurement.

The direction of wind can be measured by a rotating vane above compass bearings.

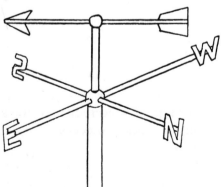

Fig. 13.6   Wind vane—showing direction of wind.

The effect of wind on crops and stock

Winds are unpleasant for stock unable to find protection. Cows give less milk, hens go off lay, and fattening or growing stock do not put on weight if exposed to a biting wind.

A strong wind, on a newly planted crop in light sandy soil, can blow the soil and seed off the field. Well-grown cereals are subject to lodging if a strong wind beats through the crop before harvest.

Windbreaks

The force of the wind can be temporarily broken by an obstacle in its path which resists its passage partially or completely.

(i) Solid barrier

These include walls or close-boarded fences. The characteristic wind-flow pattern, caused by a solid barrier across the direction of flow, is shown by fig. 13.7.

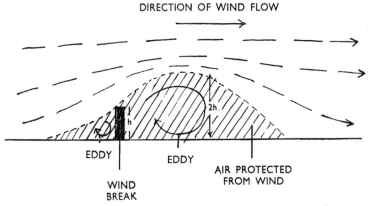

DIRECTION OF WIND FLOW

EDDY   EDDY

AIR PROTECTED
FROM WIND

WIND
BREAK

Fig. 13.7 The effect of a solid windbreak on the flow of air.

The overall result is a great wind reduction for a short distance, which is very suitable for stock protection.

(ii) Filter barriers

These include tree belts, hedges, or wattle fences. The result is to restrict the force of the wind by much less than a solid barrier but, since eddy formation is restricted by leakage of wind through the hedges etc., the effectiveness of the barrier stretches much further across the field.

Pulling out a hedge takes a short time, but growing one takes much longer. It is worth contemplating the loss of wind protection before pulling out a hedge in an exposed site.

Humidity

As air becomes heated, its capacity to hold water is increased. The amount of moisture in the air is called the humidity of the air. When air is fully saturated with water vapour, drops of liquid can remain suspended in the air and a mist or fog is thus formed. If the drops become large they will fall as rain. Clouds are suspended drops of liquid water. Fully saturated air is at dewpoint since, if it is cooled rapidly by contact with cold grass leaves or cold water surfaces, drops of water are squeezed from the air. This is dew formation. In limestone country where water for stock is scarce, many dewponds can be found. These are partly collecting rain, but also encouraging dew to form on their cold surface.

Measurement of humidity

Relative humidity is the amount of water held by the air as a percentage of the maximum which it could hold at a particular temperature. For example, 50 per cent relative humidity means that the air is half-saturated with water. An instrument for measuring moisture in the air is called an hygrometer.

(i) Natural hygrometers

Seaweed and fircones are often suspended at the door of the house to indicate the humidity of the atmosphere. The seaweed becomes swollen when damp and the fircone closes up. The seaweed dries out and the fircone opens when the weather is dry.

People and animals with curly hair find the curls becoming tighter in damp weather.

(ii) Artificial hygrometers

A twisted hair attached between two fixed points with a lightweight
pointer on the hair can indicate changes in atmospheric humidity
by twisting and untwisting. This is the principle behind some
hygrometers which measure atmospheric humidity.

These instruments can give very precise readings if properly
made and carefully calibrated. A hygrometer of this type is used to
measure the humidity of a grain sample after harvest. From the
humidity, the moisture content can be found fairly accurately.

Two thermometers side by side, one with a dry bulb and the
other with a wet one can be used as a hygrometer.

If water evaporates from the surface of the wet bulb, it will be
cooled by the removal of latent heat for the evaporation of water.
So when evaporation is occurring, the wet bulb thermometer has a
lower reading than the dry bulb. Since evaporation only takes
place when the air is unsaturated, the difference between the two
readings gives an indication of the relative humidity of the air.

The effect of humidity

In cold weather, when the air is moist, mist and fog can form. In
warm weather during the summer, humid conditions lead to the
rapid spread of fungal diseases through crops. Mildew spreads
rapidly through barley, rusts spread quickly through wheat, and
blight can spread through potatoes.

Many insect pests of crop plants and stock thrive under humid
conditions. Leatherjackets, the larvae of the crane fly, feed at the
surface when the weather is humid, and the greenbottle fly strikes
sheep when the weather is moist.

The farmer drying corn will find that air with a low relative
humidity will dry his grain well, but moist air will not dry it at all
until heated to lower its relative humidity. At least one type of
grain-drying unit refrigerates air to precipitate its moisture content
before allowing it to warm up again with a lower humidity more
suitable for grain drying.

Rainfall

When moist air is cooled by rising to a high altitude in a convection

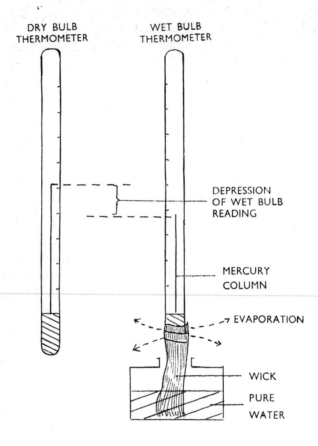

Evaporation removes latent heat from the wet bulb. The depression of the wet bulb temperature is related to the rate of evaporation which is dependent upon relative humidity.

Fig. 13.8   Wet and dry bulb hygrometer.

current or in passing above hills, the humidity reaches dew point and liquid water drops appear in the air. These drops may remain suspended in the air and move some distance as a cloud before the water is finally dropped back to earth.

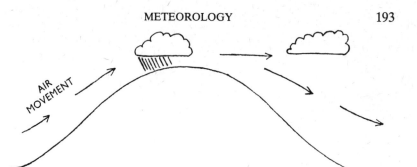

Fig. 13.9 Rain formation as air passes over hills.

Measurement of rainfall

Rain is collected in an open funnel and stored in a bottle underneath. The bottle is emptied into a suitably graduated flask which allows for the surface area of the funnel. Snow should be collected and melted before measurement since its density is very variable. The annual rainfall in the British Isles is about 1,000 mm, and slightly over in places on the western coastline bordering on the Atlantic Ocean, going down to between 500 and 750 mm on the eastern side of the country. The driest area is around and south of London.

Irrigation is used to replace rain in the drier areas of Britain, and there are occasions when its use is a marked advantage in the wetter regions. The rain from the sky is not always timely.

Although a national picture can be drawn to show rainfall figures for large areas, within those areas there are often local variations. Rain shadows occur on the lee-side (away from the prevailing wind) of large hills since the wind rises up to pass over the hill and carries the clouds over with it. Often a belt of hills causes the rain to be precipitated before the wind reaches the other side.

Evaporation

The rate of evaporation of water from a pool is dependent on atmospheric humidity, air movement, and the heat from the sun. Measurement of evaporation rate gives an indication of the speed at which plants are losing water.

If the surface of a pool were covered by felt lamp-wicks dipping into the water and projecting about 150 mm above the water, one can imagine that the speed of water evaporation would be greatly increased. The plant has an approximately similar effect, above moist soil, so the rate of plant transpiration (water evaporation from leaves) usually exceeds the rate of evaporation from open water surfaces.

In some countries where rainfall is very low, only one crop is grown in two years. In the intervening year the soil is kept bare by surface cultivation to prevent evaporation caused by plant growth. A loose surface tilth becomes dry and seems to inhibit evaporation. Laying loose organic matter on the grown surface also limits the loss of water by evaporation.

The calculation of irrigation need takes into account the rate of evaporation of water.

Irrigation need

Throughout the summer months of May to September herbaceous plants in the U.K. transpire approximately 2.5 mm of water per day.

Thus, in a thirty day month a crop would remove 75 mm of water from the soil. If during the month 75 mm of rain falls, then this deficit is made good. If there is less than 75 mm, then the soil does not hold its full quota of water. The water shortage accumulates rapidly as days go by, and should the soil moisture deficit reach 75 mm (i.e. transpiration loss exceeds rainfall by 75 mm) then many plants begin to wilt, i.e. suffer loss of turgidity because of water shortage.

If irrigation is possible, the water in the soil can be replenished to prevent the plant suffering through the drought. Even in the western counties where the rainfall is around 875 mm, during dry springs and summers the soil moisture deficit can easily accumulate to about 240 mm. Many seeds planted during such a spring fail through shortage of water. In such dry times grass can be practically non-existent for stock feed and farmers have to resort to feeding barley straw and hay in August. Farmers with irrigation equipment and sufficient water to supply their needs, can find its usefulness brings financial returns almost equal to the cost of installation.

When the soil has its full quota of water it is at 'field capacity' (see section on water in the soil in Chapter 13).

Air temperature

The temperature of the air is noticed by all men—farmers or not. The human skin is covered by sensory cells which are heat sensitive. As they are delicate and precise one notices even a slight change in air temperature.

Temperature controls humidity in the air, since warm air has a lower relative humidity. Temperature helps to control the speed of photosynthesis. Every 10°C rise in air temperature doubles the rate of photosynthesis if other factors are not limiting. After 25°C the rate is not doubled, and over 35°C the rate drops off rapidly because the plant becomes over-heated.

Air temperature will affect soil temperature also, since the air passes into the soil crevices.

Measurement of air temperature

It is usual to record the maximum temperature attained by the air during a twelve hour or twenty-four hour period, and also its minimum temperature during the same time.

The thermometer for measuring maximum temperature is an ordinary mercury column in a calibrated glass tube, but there is a steel spring which is pushed upwards by the mercury. The spring does not return when the mercury contracts, but remains at the highest point reached by the mercury column. After reading the temperature, it is a simple matter to draw the steel spring down to the mercury again with a small magnet.

The minimum thermometer is an alcohol thermometer. When the alcohol expands it easily bypasses a steel spring in the calibrated glass tube; but, on contracting, the surface tension of the alcohol draws the spring down to the lowest point reached by the alcohol. The spring is forced to remain in position until the operator returns the steel spring to the surface of the alcohol using a magnet.

An annual record of maximum and minimum air temperatures shows two things clearly:

1. A seasonal fluctuation of both maximum and minimum temperatures. High in summer and low in winter.

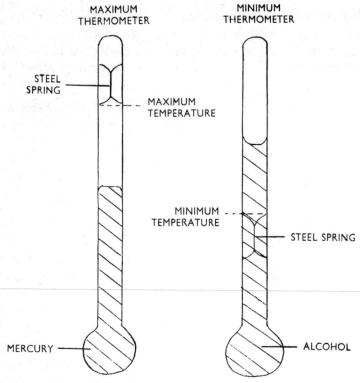

Fig. 13.10   Maximum and minimum thermometer.

2. A day-to-day fluctuation which shows a close correlation to the atmospheric pressure. When pressure is high in winter and spring, the air is cold and frosts occur at night. When pressure is low in winter, the air temperature rises.

Temperature readings used to be quoted in Fahrenheit degrees in Britain, but Celsius is now frequently used. The scale in use must be clearly stated on the record (see fig. 1.36).

Maximum and minimum thermometers are extremely useful in incubators, greenhouses, and controlled environment animal houses, as they record faithfully the lowest and highest temperatures to which the eggs, plants, or animals have been subjected.

Soil temperature

The soil acts as a blanket for plant roots and young seedlings, protecting them from sudden changes in air temperature. Thus at 300 mm down, day-to-day variation is eliminated and only seasonal variation occurs. At 150 mm depth, day-to-day variation does occur slightly but there is a delay before air temperature affects the soil temperature. Above 150 mm the daily variation becomes more obvious. Soil temperature is very important as it determines the length of a growing season, and the types of crops grown in an area.

Rye will germinate at 2.2°C, so in Britain it would germinate in January if necessary. Rye, therefore, is a very useful cereal in countries bordering on the Arctic Circle where the growing season is very short. Maize will not germinate below 8.8°C, which means that in Britain it starts growth in May. This is too late for a large cereal like maize and unless the hardier varieties can prove successful at germinating in soil temperatures lower than 8.8°C, this cereal will remain a crop for a Mediterranean climate. The soil temperature determines the effective spring and winter dates. A soil temperature of 5.5°C stimulates the active growth of grass roots, so this fixes the spring and winter dates. The usual spring date for the south-west is March 15th, and the winter date November 15th. In parts of Cornwall and Pembrokeshire the spring date is a fortnight earlier on March 1st, and winter is held back to December 1st. These fortunate counties are affected by warm winds moving with the Gulf Stream across the Atlantic. They are suited to the growth of early potatoes and spring bulbs because of their early spring.

The danger always with early growth is the chance of a delayed cold spell which could retard or kill growing plants.

Aspect

A southerly aspect allows soil to warm up faster as the sun's rays are concentrated on a smaller area (see Fig. 13.11).

Soil type

Wet soil warms slowly in the spring, so clay soil is usually late. Sand is much earlier. The specific heat of dry sand is lower than the specific heat of dry clay; this also means that sand will heat up more rapidly than clay.

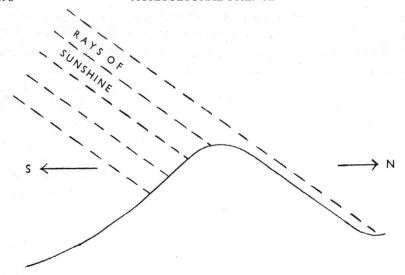

Fig. 13.11 The effect of north and south facing aspects of land on the concentration of the sun's rays.

Sunshine

The amount of direct sunshine falling on any region of the earth will depend upon the cloud cover and the season of the year.

During the summer months in the northern hemisphere of the globe the earth's axis is tilted so that the North Pole is nearer the sun than the South Pole, and the sun comes directly overhead at the Tropic of Cancer which is latitude 24° North of the Equator on June 21st.

The measurement of sunshine

The rays of the sun can be focused onto a strip of paper by using a powerful spherical lens. The sun burns a charred line across the paper which is marked off in hours. This type of recorder was designed by Campbell and Stokes: it is expensive to purchase, but simple to operate.

The effect of sunshine on the crop plant and animals.

The duration and intensity of light are important both to animals and plants. Obviously to plants the sun's radiant energy is assisting the endothermic activity of photosynthesis. There is another effect of light on plants also. Plants are stimulated to produce their inflorescence by a certain daylength, hence grasses, cereals and other plants go to seed at about the same date each year.

Animals are probably stimulated by light in a similar way. Birds begin to lay their eggs when the day length is increasing; they therefore lay their eggs in the spring. Artificially lighting the birds at night in the early winter can bring the egg laying activity of domestic fowls forward into the winter. It has been found that a short bright light stimulates them as much as a lengthy dim light. The energy output of the light is the important factor, e.g. 1,000 watts for 1 minute is equal to 50 watts for 20 minutes. The light affects the pituitary hormone gland under the brain.

Clouds

Cloud formations diffuse the sunlight falling on the earth and reflect back radiation from the sun and the earth respectively. The identification of cloud formations can help an observer to forecast local weather conditions.

To simplify the identification of cloud formations is difficult, but basically there are two groups of cloud. The cumulus on the one hand are large or small, but have a definite outline. On the other hand, the stratus cloud spreads across the sky sometimes at low, and sometimes at high, altitude.

Cumulus cloud

This cloud brings showers which can be very heavy but the rain is not continuous for a period of hours. Thunder clouds are in this group. They are so high that, viewed from the base, they appear to be very dark and ominous. Small fluffy cumulus clouds with plenty of blue sky between do not usually produce rain at all.

When cumulus clouds are dropping rain, their trailing edges become ragged and give us a brief warning of an approaching shower.

Stratus cloud

When stratus cloud is building up into sheets, it usually indicates rain in the next day or so.

Stratus rain is prolonged and can be heavy. When the sheets are beginning to break up, this is a sign that the weather is clearing up.

(B) The value of weather records

Local climates

When planning a cropping programme for a locality it is wise to study local weather records. The amount of rain which has fallen over the previous five years is a pretty good guide to future rainfall. The soil temperature record should be carefully studied, observing the date when the soil reaches a temperature of 5.5°C when grass will begin to grow. The air temperature should provide details of the earliest winter frost and the last spring frost. Rainfall figures will also indicate the amount of snow to be expected. The direction of the prevailing wind will affect the siting of new buildings.

In short, this information is invaluable if studied carefully at the outset and is best described as the climate of the area.

Correlations between records

One important correlation worth remembering is that just as pressure falls as altitude increases, so does temperature fall by 1°C for every 162 m. The Ministry of Agriculture class farms above 212 m as hill farms eligible for certain specal financial considerations. These farms are obviously affected by this temperature drop as well as their other very real problems such as inaccessibility, stony ground, exposure, and high rainfall.

When a complete annual record is studied, the pressure is found to be the key to the changes experienced.

For day-to-day weather indications at home a barometer is valuable, not as an ornament in the hall but as an important instrument to be consulted daily.

The records on page 201 were taken by the author in Gloucestershire close to the River Severn. Note the correlation between high pressure in winter and cold weather; low pressure and high rainfall with humid conditions.

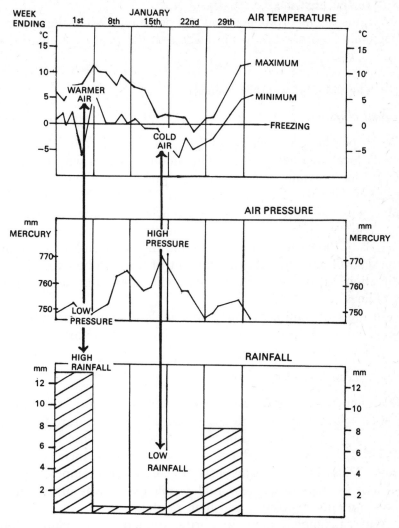

Max. temp. range
Winter: 14–0°C
Summer: 28–13°C

Min. temp. range
Winter: 12––7°C
Summer: 17–4°C

Pressure range
772–724

Soil temp. range
at 150 mm depth
Summer: 15°C
Winter: –1–+2°C

Relative humidity range

44%–100%
Av. Summer: 70%
Av. Winter: 90%
(Low lying site near to the
    river Severn)

Fig. 13.12   Copy of weather records taken at Hartpury, Glos.

(C) Interpretation of forecasts

Weather is affected by several factors such as latitude, the proximity of large land masses, sea currents, mountain ranges and large forests. Latitude is one of the most important as this affects the angle at which the sun's rays strike the ground. The latitude at the equator is 0 degrees, and the poles are 90 degrees north and south respectively. Britain is between 50 and 58 degrees north and in this position, cold polar air frequently meets with warm tropical air streams causing local turbulence.

When these air streams meet, the cold air moves under the warm air pushing it up; the warm air becomes cooled, condensation of water occurs, and rain often follows.

The leading edge of a warm air stream is called a warm front. The leading edge of cold polar air is called a cold front; and where the cold air has overtaken the warm, pushing it all to a higher level, this is called an occluded front.

A) WARM FRONT   **B) COLD FRONT**

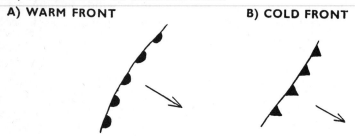

C) OCCLUDED FRONT

(WHEN COLD FRONT HAS OVERTAKEN WARM FRONT, PUSHING ALL THE WARM AIR UPWARDS)

Fig. 13.13 Symbols used to indicate fronts on a weather map.

So fronts, like depressions, are associated with rain since in both cases warm air is rising and cooling, and this causes condensation of water vapour.

Movement of fronts

Winds move fronts in an anti-clockwise direction around depressions in the northern hemisphere. Thus when a depression centres over northern Britain the fronts rotate round it, coming in from the Atlantic Ocean across the southern counties to swing northwards on the east coast. The prevailing winds are therefore wet and the west coast receives more rain than the east. Around an anti-cyclone (high pressure area) the wind moves in a clockwise direction, in the northern hemisphere. Thus when a high pressure centres on the United Kingdom, winds approach Britain from central and northern Europe. High pressure during the winter months therefore brings cold north-easterly winds lashing across the eastern counties. The month of March is often marked by cold, dry spells when a north-easterly wind prevails.

Local information is best

The national weather reports do not take into account local variation. The presence of a frost pocket is known only too well to a farmer, although not marked on a general map.

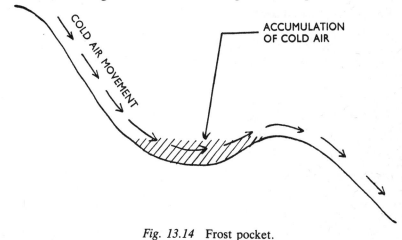

Fig. 13.14 Frost pocket.

The local weather recording stations are usually ready to help the farmer with information. Using this service, national forecasts, and local instruments, it should be possible to remove some of the chance from our reliance on the whims of the weather.

14 Soil

Geology is the study of the earth's crust and the rock formations of which it is made.

(A) Rock formation and constituent minerals

Igneous rock

The earth was formed from molten material which has solidified on the outside. The rock formed from the molten mass is called igneous rock.

Granite is a typical example of igneous rock. It contains crystals of silica in abundance. Since silica is an acidic mineral, granite is called an acidic rock.

Basalt, another igneous rock, has less silica present and more basic minerals.

Crystals of varying size can be found in igneous rocks. The depth at which the rock solidified determined the rate of cooling and therefore the sizes of crystal formed. Pumicestone, for example, was molten larva thrown onto the surface of the earth. This cooled so quickly that no crystals can be found in it.

Much of the building stone granite, such as the Princetown granite on Dartmoor, has well formed but small crystals in it. Sometimes granite contains very large crystals of silica. This is called quartz and is usually taken for rockery building etc.

The weathering of igneous rock

Water containing carbon dioxide is the primary weathering agent. Soluble salts in the rock are first to be eroded. Small pieces of rock become dislodged and travel in the water. Water with suspended particles of rock becomes abrasive and scours the rock face. Crevices appear in the rock eventually, and when ice forms in these the expansion breaks off more of the rock.

Sedimentary rock

Small particles of weathered igneous rock are carried by fast-moving streams and rivers into wide estuaries and lakes. In the slowly moving currents of the estuary the small rock particles are deposited.

Horizontal beds of sediment build up in which the particles are slowly compressed and cemented together until sedimentary rock is produced. Sometimes the sediment is composed of the shells of sea animals which produce a calcium carbonate rock like the oolitic limestone.

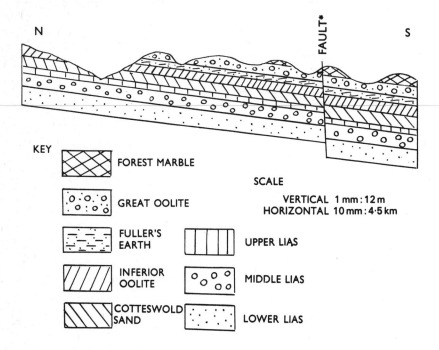

*Fault: A break in the earth's crust. The strata levels move at the fault.

Fig. 14.1 Sedimentary rock formation, a north to south section in the Cirencester area, showing sedimentary rocks of the Cotswold hills.

Some well known sedimentary rocks formed in this way are: sandstone, ironstone, limestone, chalk, and shale.

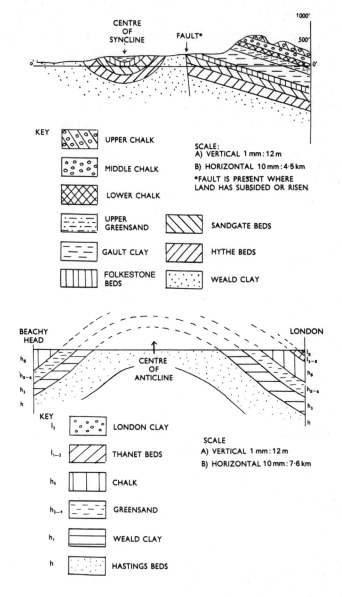

KEY

UPPER CHALK

MIDDLE CHALK

LOWER CHALK

UPPER
GREENSAND

GAULT CLAY

FOLKESTONE
BEDS

SANDGATE BEDS

HYTHE BEDS

WEALD CLAY

SCALE:
A) VERTICAL 1 mm : 12 m
B) HORIZONTAL 10 mm : 4·5 km
*FAULT IS PRESENT WHERE
LAND HAS SUBSIDED OR RISEN

KEY
l_3 LONDON CLAY

l_{1-2} THANET BEDS

h_5 CHALK

h_{2-4} GREENSAND

h_1 WEALD CLAY

h HASTINGS BEDS

SCALE
A) VERTICAL 1 mm : 12 m
B) HORIZONTAL 10 mm : 7·6 km

Fig. 14.2 Syncline; geological section from Henfield to the South Downs.
Anticline; geological section from London to Beachy Head.

Dip slopes of sedimentary rock

The molten core of the earth is continuously moving, bulging outwards in some places and sinking inwards in others. The horizontal beds of sedimentary rocks are folded by this movement of the earth's crust.

Many sedimentary rock formations can be found with steep dip slopes for this reason.

Metamorphic rock

From time to time, molten material from the core of the earth bursts upwards through the sedimentary rocks. The very great heat and pressure caused by such outbursts changes the form of sedimentary rocks nearby.

Shale, a relatively soft rock, when compressed and heated becomes slate—one of the most durable building materials known to man.

Mineral particles of rock

Mineral particles present in igneous rocks, appear again in sedimentary rocks and finally in soils.

Silica (SiO_2)

Silicon dioxide or silica is an acidic compound found abundantly in granite and, to a lesser extent, in basalt. Pure silica crystals are found in quartz. Small fragments of silica form sand.

Chemically silica is slow to react; it is used to make glass which can hold concentrated acid without being damaged. Thus a sandy soil is almost entirely without any natural supply of plant nutrients.

Alumino-silicates (clay minerals)

There are many types of alumino-silicate to be found in igneous rock. When broken up they form clay minerals.

Clay minerals consist of silicate sandwiches with fillings composed of aluminium, iron, magnesium, potassium, and sodium.

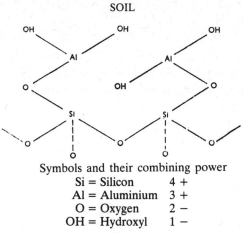

Symbols and their combining power
Si = Silicon   4 +
Al = Aluminium   3 +
O = Oxygen   2 −
OH = Hydroxyl   1 −

Fig. 14.3 Clay mineral structure. Gibbsite-simple mineral, one layer of silica and one layer of aluminium oxide.

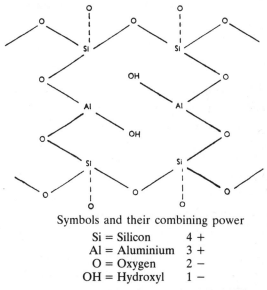

Symbols and their combining power
Si = Silicon   4 +
Al = Aluminium   3 +
O = Oxygen   2 −
OH = Hydroxyl   1 −

Fig. 14.4 Clay mineral structure. Pyrophyllite, sandwich mineral, two layers of silica and one layer of aluminium oxide in between.

Depending on the type of filling in the sandwich so the clay minerals are described as iron-rich or magnesium-rich etc. Since these clay minerals contain elements other than silicon, they are responsible for supplying the plant nutrients.

(B) Soil formation

The weathering of rock takes place over hundreds of years. The chief agent responsible is water.

Water containing dissolved carbon dioxide is acid and able to dissolve lime and other bases. Water carrying particles of sand becomes abrasive. Freezing water has the same effect as an expanding plant root, it swells in a crevice, snapping off pieces of rock. Apart from their mechanical effect, young roots also cause rock weathering by producing acids.

Soil layers

Three layers can be distinguished in most soils.
1. Topsoil supports plant growth and contains humus or decayed plant remains.
2. Subsoil is a mixture of topsoil and the underlying rock.
3. Rock.

Sedentary soil

Such a soil forms from the rock underneath it. For example, on the limestone Cotswold hills in the south-west midlands, below the thin topsoil is the brash (subsoil) which contains much limestone and below the brash is the solid limestone rock.

Transported soil

Many soils are moved by wind and water as soon as the rock is eroded. These soils form alluvial deposits elsewhere and may bear no resemblance to the underlying rock.

Activity of soil animals

The earthworm in particular mixes surface soil with subsoil in addition to physically altering the soil which passes through its body. This activity may serve to mix an alluvial deposit with the underlying rock, so forming a mixed soil which is difficult to classify.

Weight of soil per hectare

A reasonable depth of soil for agriculture is 230 mm; more than this is usually a sign of very fertile land.

A hectare of dry soil to a depth of 230 mm weighs about 3,000 tonnes.

Soil profile

A pit dug in the soil down to the parent rock reveals the changes that occur between different layers in the soil. Three zones are usually defined in a soil profile:
  1. The 'A' horizon from which minerals are leached.
  2. The 'B' horizon into which minerals are deposited from the 'A' horizon.
  3. The 'C' horizon is the parent rock.
The profile which develops is determined by an interaction of climate on rock type.

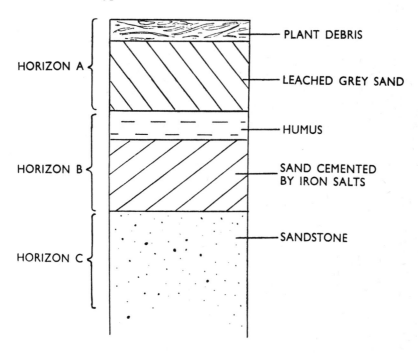

Fig. 14.5   Soil profile on sandstone (podsol).

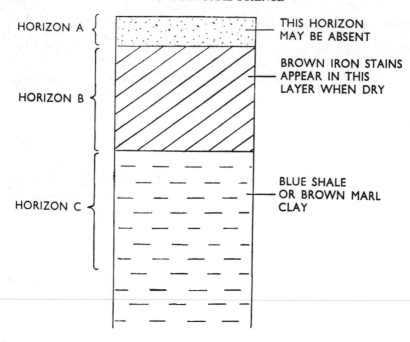

HORIZON A { THIS HORIZON
MAY BE ABSENT

HORIZON B { BROWN IRON STAINS
APPEAR IN THIS
LAYER WHEN DRY

HORIZON C { BLUE SHALE
OR BROWN MARL
CLAY

Fig. 14.6 Soil profile on clay with impeded drainage (gley).

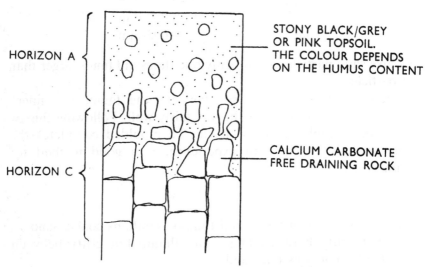

HORIZON A { STONY BLACK/GREY
OR PINK TOPSOIL.
THE COLOUR DEPENDS
ON THE HUMUS CONTENT

HORIZON C { CALCIUM CARBONATE
FREE DRAINING ROCK

Fig. 14.7 Soil profile on chalk or limestone (rendzina).

Types of soil profile

Podsol

This is found on freely drained rock, usually on sandstone. Cultivation removes the 'A' horizon which becomes mixed with humus.

Gley

This is formed when the drainage is impeded. Usually present on undrained clay. Drainage removes a gley since air can enter the soil and oxidise the green ferrous salts to the red ferric form.

Rendzina

This is found on limestone or chalk. Soils are very thin on limestone; ploughing depths are limited.

(C) Texture, structure, and classification

Soil particles are graded according to their size.

Coarse Sand	2.00–0.20	mm
Fine Sand	0.20–0.05	mm
Silt	0.05–0.002	mm
Clay	<0.002	mm

Coarse sand particles are roughly one thousand times larger than clay particles.

Rubbing a sample of dry soil between the thumb and finger indicates the composition of the soil fairly well. Following this by wetting the sample to see if it can be moulded completes the simple analysis which, with experience, is a good method for judging soil type:

Dry texture:

1. Gritty between finger and thumb—contains coarse sand.
2. Not gritty between finger and thumb but gritty between teeth—contains fine sand.
3. Silky feeling—contains much clay.

Wet texture:
1. Cannot be moulded—mostly sand.
2. Stains hand—contains some clay.
3. Can be moulded, contains over 20 per cent clay.
4. Very plastic—mostly clay.

Soil structure

Sandy soil has no structure. When it dries out, the sand particles do not adhere to one another.

Structure is important in clay soil since the particles are so small. A well flocculated clay soil has good crumb formation which leads to improved aeration. Deflocculated clay is difficult to work, and infertile because there is usually too much water in the soil.

Classification of soil types

Soils are classified on their content of clay.

0– 5 per cent clay	Sandy
5–10 per cent clay	Sandy loam
10–20 per cent clay	Loam
20–30 per cent clay	Clay loam
30–40 per cent clay	Clay
More than 40 per cent clay	Heavy clay

There are some special modifications to this simple classification:
Gravel loam contains gravel
Chalk loam contains chalk
Marl clay loam contains 5–20 per cent calcium carbonate
Calcareous soil loam contains more than 20 per cent calcium carbonate.

Descriptive soil terms

Adjectives which are often coupled with sand soil are: light, hungry, and dry. Whilst the following are linked with clay soil: heavy, mineral rich, wet, cold, sticky, tenacious, and stubborn.

Shallow soil or thin soil means that ploughing often brings subsoil to the surface.

Friable and mellow refer to a fine state of division often the result of frost action.

Natural vegetation as an indication of soil type

Poor land is often populated by bracken, whortleberry, heather, larch and pine trees.

Clay land carries oak trees well, but at ground level buttercups and sometimes cowslips are found.

Chalk or limestone carries beech woods.

Wet land is picked out easily if rushes or sedges are growing in it.

(D) Effect of soil constituents on soil fertility

Soil must supply the following five items in order to be classed as fertile:

1. Anchorage for roots
2. Water
3. Oxygen for mineral uptake
4. Major and minor nutrients
5. Protection for roots against extremes of temperature.

The soil must also be free from:

1. Poisons
2. Pests and diseases
3. High or low levels of pH.

A fertile soil consists of a balanced mixture of sand, clay, and humus so that the bad characters of sand are outweighed by good qualities in the clay and vice versa.

Clay

Clay particles are valuable in the soil as they are mineral rich. They are so small that the spaces between them are no bigger than capillaries (like small blood vessels). These small spaces do not allow water to drain readily from the soil and therefore air is prevented from entering. In dry weather, the capillaries draw water upwards from the water table much further than the corresponding water rise in sand.

Clay particles are negatively charged at their surfaces so they can hold basic minerals (cations) against leaching.

Clay has a high specific heat (0.25) which means that it heats up only four times quicker than water in the spring. The more water present in the clay, the colder the soil will become.

Clay soil provides good anchorage for plant roots and, if the structure is reasonable, drainage and aeration are improved. Clay soil cannot be cultivated when wet without damaging its structure with consequent deterioration in drainage.

Summing up the characters of clay:

Clay holds water and minerals but prevents air from reaching plant roots.

Without oxygen, plants cannot take advantage of the minerals held.

Sand

Sand soils are barren unless nutrients are supplied artificially for plant growth. The particles are large so the spaces between allow water to percolate rapidly through them. The air supply to the soil is excellent, but nutrients are leached out rapidly into the drainage water. There is very little water retained in sand soil and virtually no capillary rise from the water table. Sand is often acidic because lime is leached out.

Sand has a low specific heat (0.16) which means that it heats up six times faster than water, or half as fast again as clay in the spring. Sand soil is easily cultivated even when wet without risk of structural damage.

Summary of the characters of sand:

Sand is a warm, dry, barren soil which requires added water and minerals for satisfactory growth.

Humus

The mineralisation of animal and vegetable remains by bacteria and fungi in the soil, results in the formation of a residue called humus which is resistant to further rapid decomposition even in well aerated soil. Humus is dark in colour, not recognisable as either plant or animal remains. It usually forms a sticky colloid which coats the soil particles.

The values of humus in the soil:

Humus absorbs water so is very valuable in sand to improve moisture retention. As it is dark in colour, humus absorbs more radiant energy from the sun than light coloured soils like sands.

Humus particles are negatively charged so they can hold cations against leaching. This is almost as important in a sandy soil as moisture holding.

Humus particles become attached to clay particles causing crumbs of clay to form. This organic flocculation of clay improves its drainage and aeration.

The continued slow breakdown of humus releases a few nutrients for plant growth over a long period. Humus is particularly useful as a source of micronutrients.

Ideal soil for plant growth

If such a soil exists, it would be a loam with plenty of humus.

A loam is a mixture of sand and clay. This mixture would give the soil some large spaces for drainage and aeration while retaining many capillary size cavities or pores for water retention and capillary rise. The clay could hold the cations while the sand keeps the soil open so that oxygen can reach the roots allowing them to respire and absorb the cation nutrients.

Humus would improve the clay structure while assisting with moisture retention and cation storage.

(E) Soil improvement

Fertile soil contains millions of micro-organisms. These bacteria etc. will only be active if air is present in the soil. Water logging drives the air out of the ground. The bacteria are then limited to those which can survive without oxygen. A few survive for a short time until their waste products lower the pH to a point around 3.5–4.0 when even they can no longer survive. In the absence of bacteria, organic matter builds up on the surface. Peat formation is the result with strict limitations imposed on vegetation by the very low pH.

The improvement of infertile land should be designed to encourage the microbial population to resume their activity.

1. Air must first be allowed into the soil by drainage.
2. Lime must be liberally applied when the pH is low.
3. Green manure crops and dressings of farmyard manure should be ploughed into the soil to provide food for the bacteria, earthworms, and other useful soil animals.
4. When conditions are returning to normal, improvements in crop yield can be achieved by dressings of artificial fertilisers.

(F) Soil water, drainage and irrigation

All plants require water for growth, but they also require oxygen around their roots. If there is too much water then oxygen is pushed out of the soil. A balance of air and water in the soil produces the best plant growth, optimum conditions for earthworms, and encourages the bacteria responsible for the release of many plant nutrients in the soil.

The effect of water on soil conditions

Flooded land

Water covers the surface of the soil and fills all the pores within the ground. Plant leaves and roots are prevented from obtaining oxygen.

Waterlogged soil

There is no oxygen left in waterlogged soil as all the spaces are filled with water. Persistence of this condition will lead to the death of plant roots and the drowning out of the useful earthworms in the soil.

Many plants which are waterlogged for part of the year (particularly during the growing season), become shallow rooted and have little resistance to drought when the water level falls.

The water table is that level in the soil below which all the air spaces are filled by water.

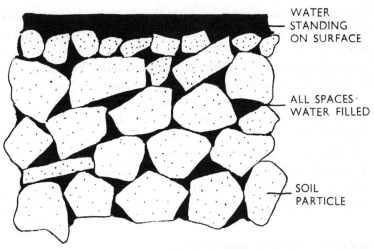

WATER STANDING ON SURFACE

ALL SPACES WATER FILLED

SOIL PARTICLE

Fig. 14.8 Flooded soil.

Field capacity

When all the larger spaces have lost gravitational water through drainage, and only the narrow spaces called capillaries retain their water against gravity, there is water and air in the soil thus providing ideal circumstances for plant growth.

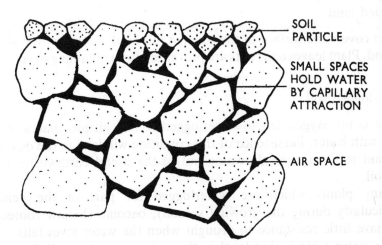

SOIL PARTICLE

SMALL SPACES HOLD WATER BY CAPILLARY ATTRACTION

AIR SPACE

Fig. 14.9 Field capacity, i.e. soil holding as much water as possible without being waterlogged.

Even when the water table falls within a soil, much of the soil above can remain at field capacity because water creeps upwards through the soil capillaries. Capillary rise is important in clay soil where upward creep of water can span several feet. In sand this capillarity does not occur to any important extent.

Wilting point

Plants begin to wilt when the available water has been drawn out of the soil. Any water left in the soil is hygroscopic and held too strongly to be extracted by roots.

Wilting point depends on the species of plant and the age of the plant. Some plants called xerophytes are drought resistant and do not wilt rapidly. The red fescue (*Festuca rubra*)—described in Section II rolls its leaves into a tube to avoid moisture loss in dry weather.

Young seedlings will survive less drought than established plants because they have not developed an extensive root system.

Summary of effect of water on soil conditions:

Too much water excludes air, so drainage is necessary to remove gravitational water.

Hygroscopic water is firmly held by humus in the soil and is unavailable to plants. In a peat soil the hygroscopic water can represent a high percentage of the total water. For this reason plants growing on peat like heather have to have xerophytic characters in order to survive.

Capillary water is available to plant roots.

Effect of soil type on water content

Sand soils are freely drained, so have an abundant air content, but the pores are mostly large so moisture holding by capillaries is restricted.

Limestone soils are often short of water because of the porous nature of the subsoil and rock.

Clay soils have plenty of capillary water available. Unfortunately drainage on clay is poor so the gravitational water often remains too long in the soil, excluding vital oxygen and resulting in the death of many plant roots.

The principles of soil drainage

The removal of gravitational water from the soil is designed to aerate the soil.

Wet soil is cold, water has a very high specific heat which means that it warms up slowly. Dry soil heats up more rapidly in the spring.

Wet clay is difficult to cultivate so drainage can increase the timeliness of planting.

Natural drainage

Soils lying over porous rock like limestone or sandstone are drained through natural underground channels.

Sloping land which is impermeable to water is drained by surface run-off during wet weather. If rain falls heavily on sloping land the surface layers may be eroded away.

Artificial drainage

Open ditches

Shallow ditches are still useful for draining much of our ridge and furrow permanent pasture. The furrows left by the ploughmen assist the natural surface drainage. This ridge and furrow is being

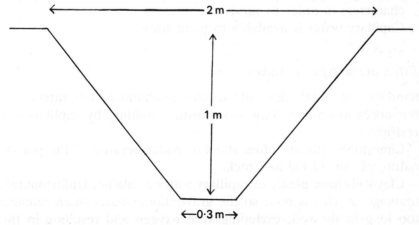

Fig. 14.10 Section through ditch on light soil.

levelled out because of the difficulty of cultivating it with modern machinery.

Many woods are drained by shallow, open ditches since tree roots block underground drain pipes.

Bank subsidence, exaggerated by stock damage, is a common cause of ditch blocking. To limit subsidence, the angle of the ditch bank on sandy soil should be at least 45° from the perpendicular. On clay soil the bank can be steeper without causing subsidence.

Collection ditches are required to gather water from underground drains. These ditches are usually sited at the edge of a field for convenience. The efficiency of the ditch drainage often determines the efficiency of expensive tile drainage layouts.

Underground drainage

Trenches dug in the ground and filled with porous material allow the gravitational water to escape from waterlogged soil. More rapid run off is achieved by laying a hollow clay or perforated plastic pipe in the trench. The loose soil above the pipe allows the water to percolate to the pipe.

Mole draining is really subsoiling, that is breaking through the plough pan, to allow the water to reach more permeable layers of subsoil beneath. The mole share is pulled through the soil at a depth of about 600 mm; the torpedo-shaped base smooths out a channel in the clay subsoil which can remain open for 6 to 15 years.

Mole drainage has the advantage that it is less expensive than piping, but its chief disadvantages are that:

1. it is parallel to the surface of the ground, and,
2. it breaks down quickly on sandy loam.

Drains laid too deeply either receive no water or lower the water table too greatly on free draining soil.

Drains which are too shallow are damaged by cultivations and surface traffic.

The principles of irrigation

After the winter rain most soils have reached field capacity. When growth commences in March or April, water is extracted from the

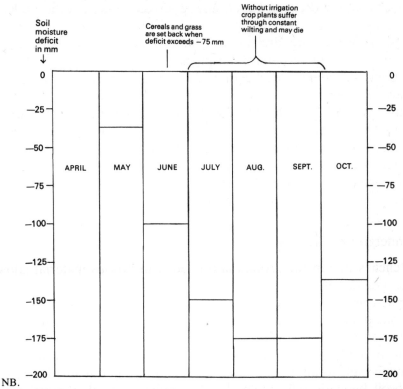

Fig. 14.11   Accumulating soil moisture deficit in eastern countries.

NB.
a) 0 = Field capacity, i.e. soil holding as much water as possible without being water-logged.
b) Rainfall in April and September is equal to transpiration loss, so the deficit is not increased during these months.

soil by plants. The rainfall in part replaces this removal of water, but as summer progresses the rate of plant transpiration speeds up and rainfall is reduced. A soil moisture deficit builds up and this becomes worse each month until the September and October rains arrive and the plants stop growing again.

The factors which control the soil moisture deficit are rainfall and the transpiration rate of plants. When the deficit reaches –75 mm, shallow rooted plants—like some grasses—begin to suffer and wilt.

When crops are irrigated, the soil should be returned to field capacity. The irrigation requirement is therefore equal to the soil moisture deficit.

Month	Rainfall mm	Estimated potential transpiration mm	Monthly soil moisture deficit mm	Cumulative soil moisture deficit mm
Mar.				
April	37	45	−8	−8
May	50	83	−33	−41
June	56	70	−14	−55
July	18	87	−69	−124
Aug.	25	93	−68	−192
Sept.	18	50	−32	−224
Total	204	403		

Take estimated potential transpiration from monthly rainfall to obtain monthly soil moisture deficit.

14.12 Table showing how soil moisture deficit was calculated in Gloucestershire in a recent dry summer.

The quantity of water required for irrigation

10 tonnes of water cover 1 hectare to a depth of 1 mm approximately.

$$1 \text{ tonne of water} = 1{,}000 \text{ litres}$$
$$1 \text{ hectare mm of water} = 1{,}000 \times 10 = 10{,}000 \text{ litres}$$

This approximate figure of 10,000 litres per mm of water makes calculation of water requirement easier.

Access to a river will solve water supply problems for irrigation, but streams are apt to dry up during dry weather.

Reservoirs are necessary when a stream or spring is the sole source of irrigation water.

A reservoir can be filled during the winter months.

Some expert advice should be sought when considering the building of a dam.

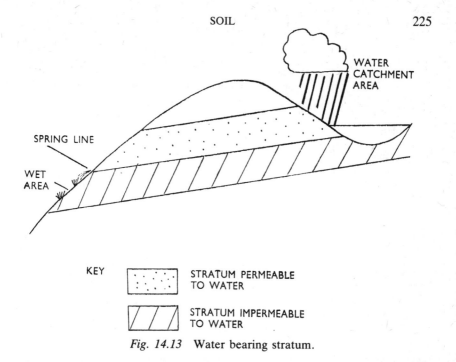

Fig. 14.13   Water bearing stratum.

(G) Natural water supplies

The water cycle

Pure water evaporates from the sea and large inland lakes. In warm air this vapour rises and is blown inland. As the warm air rises over high ground, it is cooled and the relative humidity reaches dew point. Water then falls as rain or snow. Some of this water runs off the surface of the hill into streams, rivers, and back to the sea. The rest soaks into the porous strata of the hill to replenish the water table.

The water which is stored in these underground porous rocks is often an important source for drinking purposes on outlying farms, and can be useful for irrigation if the supply can be maintained.

The important water-bearing rocks are sandstone, limestone and chalk when these are found above an impermeable stratum.

If the pressure of water is sufficiently high in surrounding hills, then a borehole will bring water to the surface without pumping. For irrigation purposes a flow of 45,000 litres per hour is desirable.

Expert advice will be necessary before drilling is carried out for

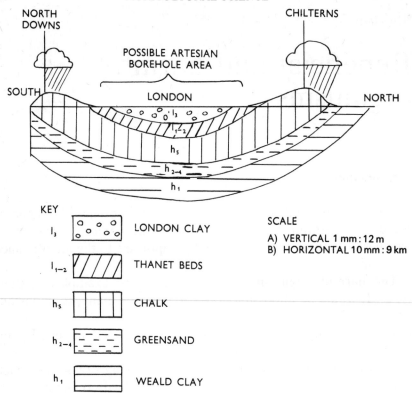

NORTH
DOWNS

CHILTERNS

POSSIBLE ARTESIAN
BOREHOLE AREA

SOUTH   LONDON   NORTH

l_3

l_{1-2}

h_s

h_{2-4}

h_1

KEY

l_3 LONDON CLAY

l_{1-2} THANET BEDS

h_s CHALK

h_{2-4} GREENSAND

h_1 WEALD CLAY

SCALE
A) VERTICAL 1 mm : 12 m
B) HORIZONTAL 10 mm : 9 km

Fig. 14.14   Artesian borehole, geological section N–S through London showing water catchment area and possible position of artesian wells.

water. A knowledge of the underlying geological strata is essential before advice on possible water production can be given.

Controls are now being imposed on those people extracting water from rivers or the water table, to ensure a fair share for all and to avoid water wastage.

Section V

Beneficial and harmful organisms

Introduction

The soil is alive with small animals and micro-organisms. Some of these are harmful to our crops, but the majority are either harmless creatures or they are vitally important to the maintenance of fertility.

The harmful organisms described here are examples of their types. It would be impossible to try to describe many, so important ones have been chosen.

The harmless and beneficial are not described at length, but their valuable activity is discussed.

A farmer inherits with his land, all the living organisms which the soil contains. While he may be concerned with removing a few harmful pests, he must also attempt to maintain the beneficial soil population by the correct care and treatment of his ground.

15 Microbiology and principles of hygiene

Microbiology is the study of those organisms which are too small to be seen without the aid of a powerful microscope. These organisms, however, play an important role in the maintenance of soil fertility, the fermentation of silage, the spoilage of stored grain or hay, the diseases attacking plants and animals, and the digestion of green food by grass-eating animals. Their occurrence is world wide. Wherever moist air and warmth come together micro-organisms can appear and, supplied with suitable food, they will develop rapidly.

The organisms mentioned in this chapter are examples of their type, it would be impossible to mention more than a few in this short space.

Major groups of micro-organisms

The micro-organisms responsible for many of the important biochemical changes connected with agriculture are divided into three groups:
1. Fungi
2. Bacteria
3. Viruses

The fungi are usually the largest organisms. Parts of many of them can be seen with the naked eye. Bacteria and viruses are smaller. Bacteria can just be seen under a powerful optical microscope, but a virus needs something with more powerful magnification, like the electron microscope, to detect it.

(A) Fungi

(a) Structure and reproduction

The fungi are plants without any green pigment for photosynthesis. They obtain all their food in a complex chemical form as do animals from other green plants. The majority of fungi obtain their food from dead plant or animal remains; but some will attack, and feed at the expense of a living host, e.g. the potato blight or a ringworm on cattle.

The plant form is simple; in most cases it is a mass of interwoven fibres called hyphae which form a mass called a mycelium.

The hyphae are made of hollow branching tubes.

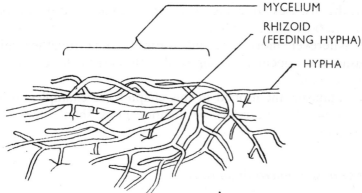

MYCELIUM

RHIZOID
(FEEDING HYPHA)

HYPHA

Fig. 15.1   Hyphae and simple mycelium of fungus.

Special feeding hyphae branch into the food, digest it with enzymes and extract its nutrients.

Reproduction in a fungus can be very rapid. When temperature, humidity, and nutrient supplies are at their best, multiplication and spread of the fungus occurs.

The most common method of propagation is by spore production. The spores arise either directly from the end of a growing hypha, or within a special container called a sporangium.

(b) Saprophytes

Saprophytes are plants which feed on dead or decaying organic matter. The food spoilage fungi come within this category.

A) SPORANGIUM WITH SPORES

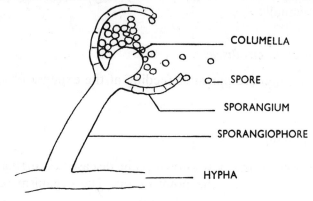

B) SPORE FORMATION DIRECTLY FROM HYPHA

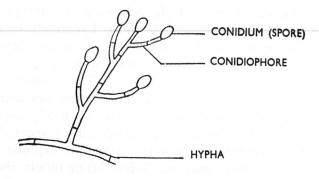

Fig. 15.2 Spore formation.

Moulds are very simple fungi; their spores are blowing about in the air. When these spores land on damp food suitable for their growth, they germinate to form a whitish mass of hyphae which causes the spoilage of stored products.

Grain, hay and straw are all affected by mould if stored damp in contact with fresh air. The mould digests some of the carbohydrate food present in the grain or hay, lowering its feeding value. The waste products of many fungi are unpleasant to the taste and even poisonous, especially to certain types of bacteria. These waste products can contain antibiotics which can upset the bacteria in the rumen if they are eaten by a cow. The animal is then unable to digest grass properly.

Although simple fungi, moulds cause a large wastage of stored food annually.

(c) Plant fungus diseases

Disease fungi are parasites feeding at the expense of a living plant host.

Potato blight

This is one of the best known plant diseases occurring every year, to some extent, on the potato crop. The organism responsible is *Phytophthora infestans*, a fungus.

The disease is brought into a crop by an infected tuber. The plant forming from this tuber is blighted.

A mycelium develops between the cells within the leaf, pushing feeding hyphae or haustoria into the cells. Eventually sporangia are pushed out of the stomata and spores are blown away. On landing, the contents of these spores form motile zoospores which can swim through a moist film on the leaf. The zoospores enter fresh leaves of potato through the stomata, thereby spreading the infection. The whole of the plant above ground is finally blackened by the fungus. Drops of water, containing active zoospores, will fall upon the ground frequently; if any tubers are exposed above the soil, they will become infected. Before lifting tubers, the tops can be destroyed by spraying with sulphuric acid. Two weeks later the tubers can be harvested with less fear of infection.

Control of the disease, or better still prevention, can be effected by spraying with copper or zinc compounds when the disease is likely to spread.

Beaumont periods, when blight will spread, are forecast to farmers. (Beaumont period: relative humidity 75 per cent and temperature over 10°C for forty-eight hours.)

The best preventative measure is to adopt a regular three-weekly spraying programme from June onwards. Protection of the tubers is ensured by earthing up the plants adequately.

A programme of plant breeding designed to incorporate the resistance to blight of the wild South American potato into a commercial plant has so far had limited success. Some varieties are

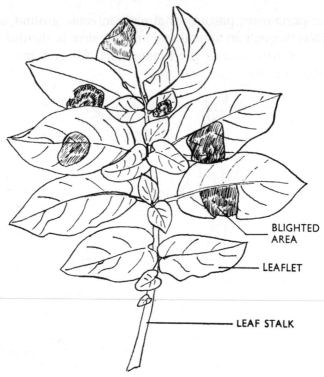

BLIGHTED
AREA

LEAFLET

LEAF STALK

Fig. 15.3 Blight infected potato leaf.

more resistant to blight than others but none has proved to be
totally resistant yet.

Club root of the brassicae

One of the troublesome diseases of the cabbage family is club root
or finger and toe disease. It is caused by a fungus called
Plasmodiophora brassicae.

Germinating resistant spores enter the root hairs of young
brassica plants. Inside the cell they form a multinucleate mass
which can release more zoospores into the soil. The damaged host
root forms large swellings or galls round the infection. The
resistant spores which are present in the galls are released when
the gall is broken. As the spores can survive in the soil for up to
seven years it is important not to grow brassicae again on the same
land too soon. The diseased swedes, turnips, or kale should be fed

on either permanent pasture or already infected ground, since the spores pass through an animal and are still alive in the dung. Sandy soil is more favourable to the fungus than clay soil and a low pH encourages the disease.

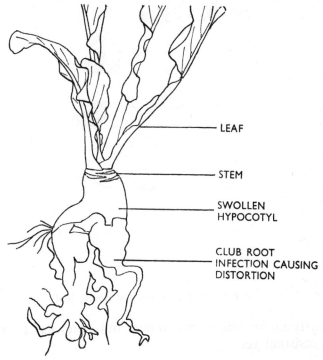

Fig. 15.4   Swede infected with club root fungus.

Diseases of cereals

Fungal diseases of the cereal crop are many. The three ways by which they are spread are:
1. Wind-borne
2. Seed-borne
3. Soil-borne

An example of a wind-borne disease is yellow rust of wheat. This is caused by a fungus called *Puccinia striiformis*. The spores are resistant in the soil over winter. During the spring, spores are formed which are blown onto the growing crop by the wind. The mycelium develops on the surface of the leaf, penetrating the leaf cells and destroying chloroplasts. The spores form on the leaf surface which is then coloured yellow. Too much leaf growth on a cereal in spring encourages the disease.

An example of a seed-borne disease is covered smut of wheat (Bunt). This disease of cereals is largely disappearing, because satisfactory preventative measures have been adopted by merchants who dress the seed with fungicide before it is sent to farms.

The fungus responsible is *Tilletia caries*.

The spore infects the germinating grain and the fungus continues to develop inside the plant. The ripe grain is full of smut spores which are released as the grain is threshed. Infection occurs in the threshing drum as the loose spores adhere to the outside of healthy grains.

Organo-mercurial seed dressings protect the grain even if smut spores are present because as the smut germinates it is poisoned.

Another seed-borne disease is loose smut of wheat and barley. The fungus causing these diseases is called *Ustilago nuda*.

The spores of the fungus infect the developing ear so grain is contaminated before it is fully formed.

Infected seed looks healthy, although the mycelium might be detected by careful staining and examination under the microscope. Infected seed when planted may grow to form several tillers, most of which have early maturing ears completely blackened by smut spores. The spores blow onto developing grain around.

Some varieties of wheat are more resistant to loose smut disease than others, but this situation can be suddenly altered by a change in the smut. The level of loose smut in a crop of seed wheat is estimated by a field inspection. If the crop falls below accepted E.E.C. standards it is rejected for use as seed.

Soil-borne diseases accumulate on crop residues, in soil under continuous cereal cropping: take-all, eyespot and, more recently, leaf blotch are three important examples.

Take-all or Whiteheads disease is caused by a fungus called *Ophiobolus graminis*.

Symptoms of attacks in cereal crops could be irregular patches of stunted plants or shrivelled, prematurely ripened ears. When an affected plant is pulled out of the ground, its roots are only about half-an-inch long. The base of the stem is blackened by infection. This can be seen by stripping the basal leaf sheath away from the plant.

Infection arises from a persistent mycelium which can overwinter in the soil on decaying stubble, couch grass, or certain other stoloniferous weeds. As the fungus develops within the host cereal plant, ascospores are formed on the outside of the leaves. These spores die in the soil. Antibiotics produced by other fungi prove too powerful for them.

Light soil seems to favour the disease, but it is by no means restricted to sand soil. A three year rotational break from wheat and barley growing practically finishes take-all as long as grass weeds are controlled.

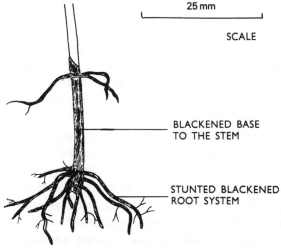

25 mm

SCALE

BLACKENED BASE TO THE STEM

STUNTED BLACKENED ROOT SYSTEM

Fig. 15.5   Wheat plant affected by 'take-all'.

A high concentration of carbon dioxide in the soil inhibits the growth of take-all, so compaction of a seed bed and adequate liming both build up carbon dioxide in soil and diminish take-all. Oats seem to be relatively immune from take-all and may help to reduce danger of an unhealthy build-up if introduced as an alternative cereal crop.

Eyespot is caused by *Cercosporella herpotrichoides*, a fungus. The symptoms of eyespot are patches of lodged plants (plants dropping over at ground level) with no general direction of fall. When the plants are examined closely a blackened area is seen at the base of the stem. This sometimes resembles an eye in shape.

The control measures for eyespot are similar to those for take-all, i.e. a rotational break from continuous cereal cropping. Certain varieties of wheat have a greater resistance to eyespot than others. The National Institute of Agricultural Botany publishes variety lists annually, showing disease resistance where it exists. Fungicides are widely used to control the level of eyespot infection when wheat is planted for a second or third successive year on the same land, but varieties showing greatest resistance to the disease should be planted wherever possible.

(B) Bacteria and viruses

Bacteria are microscopic unicellular organisms, mostly free living in the soil as saprophytes obtaining their food as do fungi by digesting complex animal or vegetable remains. Many bacteria are able to live within the body of a plant or animal; some are even obliged to live thus.

Bacteria are classified into groups by their shape reproductive habits, and their growth requirements. Further identification is carried out by observing their reaction to certain organic dyes.

They are more easily seen and studied under an electron microscope than under an optical one because of their small size.

Although bacteria are associated with disease, not all of them are harmful; indeed, without their activity agriculture would be drastically altered if it did not become impossible.

(a) Bacteria and disease

Parasites

Not all micro-organisms growing within animals or plants can be termed parasites. Those living in the rumen of a cow are beneficial; others always seem to be present but do no apparent harm or good. Many are feeding at the expense of the host animal, however, and if they provide nothing useful in return they must be termed parasites.

Pathogens

When these microbial parasites invade an animal they penetrate its body through the skin, the lining of the lungs, or the lining of the gut. If they manage to overcome the natural resistance of the body to attack, they become lodged in the tissue. Some bacteria form local lesions of infection; others migrate about the body causing widespread infection.

Toxins

Poisons called toxins are released from the pathogenic bacterial cell as it disintegrates. Some cells may begin to lose toxin before disintegration.

Toxins like other poisons interfere with the correct functioning of body processes, so preventing the animal from living normally. The reaction of the body to this interference causes visible changes in an animal's behaviour recognisable possibly as the symptoms of a disease.

Diseases

Many diseases of livestock are caused by pathogenic bacteria. Some diseases are confined to particular species of livestock, while others affect many animals including man.

When outbreaks of serious diseases occur, the Ministry of Agriculture veterinary surgeons take control in an attempt to avoid any unnecessary spread of the causative organism. Restrictions on the movement of livestock are imposed when outbreaks of 'notifiable diseases' occur.

Some diseases like tetanus, which can rapidly be fatal, linger in the soil by forming resistant spores. Farmers must therefore be on the look-out for cuts on livestock through which the bacterium could enter. Cleaning and disinfection of wounds is an important step in avoiding bacterial infection.

The udder of a dairy cow seems to be an ideal environment for bacterial growth. An udder disease called mastitis, has been the scourge of dairy herds for many years. The various bacteria which cause udder inflamation can be spread fairly easily at milking time.

Until the introduction of antibiotics, the most important bacterium causing mastitis was called *Streptococcus agalactiae*. Although streptococcal mastitis has been largely controlled by antibiotics, unfortunately other more resistant bacteria often take its place. Research seems to indicate that strains of bacteria may be developing which are resistant to certain antibiotics. Control is still possible at present by changing the type of antibiotic used to control clinical infections.

Two important diseases of livestock can be transmitted to man through milk. One is bovine tuberculosis, the other is brucellosis. These diseases have now been controlled by a national programme to eliminate infected animals from dairy farms.

Apart from these two diseases of cattle, milk can carry many serious diseases of man should it become contaminated. Diphtheria, typhoid, and dysentery are three examples. Cleanliness throughout the milking and milk handling processes is very important. Pasteurisation or sterilisation is the only way of being fairly sure that no germs have entered the milk bottle.

(b) Immunity

Natural resistance to a disease is present to some extent in all animals. The vital stage in the battle against an infection is at the beginning, when the pathogenic organism is trying to lodge itself in the host tissues.

The white corpuscles in the blood are very important at this time. Some of them engulf bacteria and render them harmless, while others produce chemicals which inactivate the toxins of the bacteria or the bacteria themselves.

Passive immunity

This implies that an animal receives help in the fight against a disease.

Firstly, this help may be given naturally by a cow to its calf as the latter drinks its dam's milk. The colostrum of a mammal, or the first one or two days' milk supply, is very rich in disease-fighting chemicals called antibodies or anti-toxins. These antibodies inactivate the toxins of specific bacteria. There is a special antibody produced to deal with each different toxin. A calf should always be allowed to drink the colostrum produced by a cow when newly calved, even if after that it is weaned on to a different diet. The valuable antibodies present in colostrum help to protect it in its first weeks of life.

Secondly, passive immunity can be conferred artificially. If the toxin can be removed from a colony of active bacteria and injected, in a harmless form, into a healthy animal the healthy animal produces antibodies in its blood which combat the toxin. Some of the blood liquid, or serum, containing these antibodies can then be transfused into the blood of the animal suffering from the disease.

This immediate supply of antibodies has helped many animals to overcome a disease by helping them in the early stages until they can produce antibodies of their own.

Active immunity

This means that an animal has protected itself against a disease by its own efforts.

Recovery from a disease usually confers a temporary or lasting immunity on the sufferer. The actual antibody or the ability to produce antibodies lingers in the blood.

Artificial stimulation to produce antibodies without the animal actually having a disease can be achieved in some cases by injections of small amounts of inactivated toxin or weakened, dormant bacteria. The animal's defence mechanism is stimulated to produce antibodies which protect the animal should it be invaded by the particular pathogen concerned.

(c) Antibiotics

In recent years antibiotics or chemotherapeutic substances have saved thousands of people from death and suffering.

The discovery of these substances is largely attributable to Sir Alexander Fleming, a British scientist. In 1929 he discovered that penicillium mould killed a colony of staphylococcus bacteria. He went on to demonstrate the bactericidal properties of penicillin, the antibiotic substance extracted from the growing mould, and also that it did not harm animals.

In 1939 British scientists began work on penicillin production. War threatened to interrupt the programme, so the work was taken to America. In three years, commercial production was possible. Its life-saving properties before the end of the war alone, justified its production. Penicillin does not kill all bacteria and searches were, therefore carried out for other antibiotics. Streptomycin was the next to be found, and is (like penicillin) produced by a fungus.

Teat injections of antibiotic have become the standard treatment for dairy cows suffering from mastitis, infection of the udder. Prevention is always better than cure and cheaper in the long run; so when high yielding dairy cows are 'drying-off' after a lactation, they are often given injections of pencillin into the udder to prevent infection.

The inclusion of small amounts of antibiotic in the diet of pigs and poultry produced improvements in growth rate. This was probably because the antibiotic reduced the number of intestinal bacterial parasites which tended to limit growth rate. This practice could prove to be dangerous to man in the long term as disease-causing bacteria may develop a resistance to certain antibiotics which are used in human medicine.

A ruminant could not afford to swallow any antibiotics as they would kill the bacteria in its rumen which are responsible for the digestion of grass.

(d) Viruses

These micro-organisms are extremely small, often passing through special filters which hold back bacteria.

They are known to be responsible for some animal disease such as foot-and-mouth disease and virus pneumonia. They also cause plant diseases; for example, sugar beet yellows, leaf roll, and mosaic of potatoes.

The spread of many plant virus diseases is effected by insects which we call vectors or carriers of the disease. The most important group of insect vectors is undoubtedly the aphid family. A green aphid spreads leaf roll and mosaic on potatoes, and a black aphid spreads much of the sugar beet virus yellows.

Virus infection seems to cause a plant reaction or symptoms characteristic of a particular disease. So in this respect their effect is similar to tha produced by bacterial pathogens.

(C) Spoilage and storage of plant products on farms

When plant products are stored alive they continue to respire, producing heat, carbon dioxide, and water. If air is excluded from the plants stored, respiration continues anaerobically producing heat, carbon dioxide, and alcohol or acid.

Continued respiration reduces the feeding value of the plants in store, as the carbohydrates are used up. The heat produced by respiration encourages the growth of bacteria and fungus which can cause rapid deterioration in a store. The effect of frost is serious on some things like potatoes. Evaporation rapidly reduces the water content of fresh greens, so products like lettuces or winter greens need careful handling.

(a) Prevention of spoilage

The prevention of spoilage is summed up in these four aims:
1. Limit respiration and growth
2. Limit evaporation
3. Insulate against frost
4. Prevent infection by bacteria, fungi, or pests.

The limitation of respiration and growth involves different techniques with different crops. Listed below are most of the methods used:
1. Reduce temperature (cold storage)
2. Remove oxygen (ensiling)
3. Allow carbon dioxide to build up (sealed silos)
4. Dry the atmosphere (enzymes require water before they are active)
5. Keep in darkness (necessary for potato but not for all plant products)
6. Add growth preventing vapours (anti-chitting vapours for potatoes)

Prevention of evaporation is usually done by deep freezing where plants are for human food.

Potatoes, mangolds, and swedes all lose water in store. Mature potato tubers seem to be less susceptible to water loss than new potatoes.

Insulation against frost can be effected by lining a building with straw bales or covering an outdoor clamp with straw. The straw contains still air which is an excellent heat insulator.

Preventing fungus or bacteria entering and infecting a food store is extremely difficult. Attempts to prevent any diseased material entering are usually worthwhile. Thorough fumigation of the store before filling may also be worthwhile.

Insect pests like grain weevils cause heating in stored grain. Drying the batch again and fumigating the bin helps to arrest such attacks.

(b) The preservation of silage

Grass is covered with thousands of micro-organisms of different kinds. If this grass is cut and packed into a tight pile to squeeze out the air, several kinds of bacteria can grow. They feed on the juice escaping from the grass, and as they respire anaerobically they produce heat. The heat builds up in the whole mass and, with the increased temperature, the products of bacterial growth accumulate rapidly.

Under good conditions, lactobacilli are the bacteria which develop. These are rod-shaped lactic acid forming bacteria. With plenty of available sugars in the plant sap these bacteria will produce sufficient lactic acid as a waste product, in about three weeks, to lower the pH to below 4 (i.e. make the medium very acid). This acid is palatable to livestock but it prevents all further growth in the silage which remains pickled until used.

Young green material may contain insufficient available sugar for the lactobacilli. Older grass has plenty of sugar in it but the protein content is low and the indigestible fibre is high. Chopping young grass helps the bacteria to obtain their sugars more readily.

If air is not excluded from the clamp, respiration continues in the grass causing the temperature to shoot up to 50°C and over which damages the protein and can produce ammonia which damages or destroys the silage.

Excess water and soil on the grass ensiled allows clostridia, butyric acid forming bacteria, to develop. Once these organisms begin growth, the butyric acid taints the silage reducing palatability.

As long as a well made clamp remains sealed from the air and protected from the rain it will keep for more than one season if necessary.

If rain washes the acid out of the silage, putrefying bacteria begin to grow giving the silage an unpleasant smell. Moulds can also grow quickly on silage if the air is allowed to penetrate.

Some farmers add chemicals to the ensiled green material to help the correct fermentation to occur. These vary from simple additives, like molasses, said to provide available sugar to encourage the growth of *Lactobacillus*, to more complex ones which help fermentation in several ways, and finally to those which contain organic or inorganic acid to speed the acidification process.

Such additives, properly used, probably pay for themselves by ensuring optimum conditions for fermentation. Many people make good quality silage without additives, but many more make poor silage without really knowing the reason for their failure.

(c) The preservation of milk

Milk is the special food produced by a female mammal to feed her offspring early in life.

Milk contains carbohydrate, protein, fat, vitamins, minerals, and water. It is therefore an ideal medium for bacterial growth, especially as it is produced from the mammary gland at blood temperature.

Milk spoils rapidly, so standards of cleanliness must be high in the milking parlour and dairy.

As cows are being milked, lactobacillus bacteria contaminate the milk. In hot weather, within a matter of an hour or so, lactic acid produced by the growing bacteria curdles the protein in milk and the acid imparts a sour taste to the liquid.

Cooling the milk down to 5°C slows down the souring process. Keeping the milk in a tank at 5°C is better still.

Dairies which receive farm milk supplies carry out regular tests to maintain high standards for their produce.

Keeping quality and bacterial tests on raw milk

The old keeping-quality test for raw milk was based on the taste and smell of milk at six-hourly intervals. The smell of milk is still checked as an early indicator of souring before a tanker driver empties a farm bulk tank.

Chemical dye reduction tests were used for many years as standard indicators of keeping quality. The speed with which oxygen was removed from certain chemical dyes indicated the bacterial concentration in the milk. The dye changed colour as it was reduced, indicating the spoilage of the milk.

More recently, the bacterial content of milk has been estimated by growing all the organisms from a small milk sample on a nutritive jelly and counting the number of colonies which result.

Theoretically every bacterium or clump of bacteria will develop to form a colony when incubated at 37°C for forty-eight hours on a nutrient agar jelly. These colonies are visible to the naked eye so are fairly easy to count.

The milk under test is diluted with water, then mixed with agar jelly in a petri dish and incubated. After two days colonies of bacteria will have formed and if the amount of milk present is known then the numbers of bacteria present can be estimated.

Plate counts not only indicate fairly accurately the likely numbers of bacteria present, but they also provide a chance for an experienced operator to identify some of the troublesome bacteria. Possible sources of contamination can then be suggested to the farmer.

Heat treatment to improve keeping quality and cleanliness of milk

Apart from the milk-souring bacteria, other bacteria may be present in milk which can cause diseases in man.

Boiling milk kills all bacteria but destroys some of the flavour. This technique is used for milk sterilisation.

A modification of Pasteur's wine preservation technique is used to kill bacteria in milk without destroying its flavour. If the milk is protected after pasteurisation it will keep much better.

Heating milk to 62.8°C for thirty minutes gives effective pasteurisation. Another commercial technique called the 'High Temperature Short Time' (H.T.S.T.) method is common, when milk is heated to 71.7°C for fifteen seconds.

Slight deviations in either time or temperature on either of these techniques can result in some bacteria escaping the heat treatment.

Milk is pasteurised to kill possible pathogenic bacteria as well as lactobacilli. It was noticed that the enzyme phosphatase, present in milk, is fairly resistant to heating. It only disappears after efficient pasteurisation.

Milk which has not been pasteurised properly could still contain phosphatase, as well as pathogenic bacteria of course. If phosphatase is detected chemically in milk after pasteurisation, the milk must be re-pasteurised.

(D) Hygiene

Cleanliness in milk production is obviously very important. It is not sufficient for milking equipment just to look clean, it should also be free from bacterial growth.

(a) Use of swabs and rinses
Swabs

The presence of bacteria on fixed milking or milkhandling equipment can be detected by wiping a sterile cottonwool swab over an area of one square foot. The swab and the bacteria which it probably contains are dropped into nutrient solution. Drops of this solution are placed on agar plates and incubated. The numbers of colonies produced after forty-eight hours gives an estimate of contamination on the surface tested.

Rinses

Rinsing a churn or milk bottle with sterile water removes some of the contaminating bacteria. These can then be counted by plating drops of the liquid rinse with nutrient agar and counting the bacterial colonies after two days.

Tests like these are important when checking the efficiency of cleaning in a dairy. Dirty containers can quickly cause milk to sour.

Washing

Soaps and detergents dissolve particles of fat which are then dispersed in the washing water.

A thick coating of fat will prevent chemical disinfectants from doing their work.

(b) Disinfection and sterilisation

Disinfection

The aim is to kill potentially harmful micro-organisms together with those which will cause spoilage. Sterilisation is desirable but seldom achieved.

Physical sterilisation

Objects placed in a hot air oven at 170°C for ninety minutes are effectively bacteria-free when removed. Such objects should be wrapped in paper to prevent contamination on removal.

Autoclaving or pressure cooking is satisfactory for equipment which can stand boiling temperatures and which is not too large to go into the autoclave. Pasteur designed the first autoclave in 1885. A pressure of 1.05 kg/cm^2 applied to the lid of a closed container in which water is boiling can raise the temperature to 120°C. This temperature more quickly kills heat-resistant bacteria.

Chemical sterilisation

It is important to use chemicals for sterilising where equipment cannot be dismantled for cleaning or when, for other reasons, heat sterilisation is not possible.

Alcohol is a desiccant, removing water from cells, so can be used to kill bacteria. Before injections are administered to animals the skin should be cleaned with alcohol.

Phenol, also known as carbolic acid, in a dilute form destroys bacteria without harming animal tissue. Phenol is the basis of most liquid disinfectants which could be used to sterilise skin wounds.

Sodium hypochlorite breaks down readily to produce chlorine, an acidic gas, which is soluble in water. Chlorine kills germs while in dilute form it does not harm humans. Sodium hypochlorite is therefore acceptable as a cleanser for dairy utensils. Iodine is a similar substance to chlorine; it is a useful clinical disinfectant for skin wounds. An iodine compound solution is used as a dairy utensil cleanser which can be conveniently sprayed on surfaces to be cleaned.

Buildings used by stock can be cleaned and disinfected to some extent by spraying internal walls with lime-wash. The lime is in the form of calcium hydroxide which being an alkali is caustic and bactericidal. Care is necessary when using lime-wash as it is caustic to the skin and dangerous if it enters the eye.

Creosote is a useful fungus preventative when applied to wood. It is not possible to paint over creosote, but where wood is exposed to water or buried in the soil some protection against rot is desirable.

(c) Sources of contamination

Aware of the dangers lurking in dirt and the risk of contamination from the air, it is disturbing to discover that water used for washing may itself be a dangerous source of bacterial contamination.

The disposal of human and animal sewerage has in the past left much to be desired. As a result many wells supplying water have been contaminated by faecal bacteria (bacteria present in dung).

The presence of coliform bacteria in water is enough for the Public Health authorities to declare a well unfit for supplying drinking water. Coliform bacteria are intestinal dwellers which are fairly harmless to adult animals; but, if they are present, typhoid, dysentery, and other dangerous pathogens may be present also.

Coliform test

A test for faecal contamination can be carried out on water and milk. Positive results show that the liquid is obviously unfit for human consumption.

Sewerage disposal

Public sewerage disposal works separate solid from liquid first. Solid waste is dried and then burned. Liquid is sprinkled onto gravel where small plants and animals grow. These feed on the micro-organisms and nutrients in the liquid. Further filtration and sedimentation of the liquid should clear from it all traces of contamination.

Septic tanks installed on many farms and in country districts are pits in which small plants and organisms develop to break down the waste into harmless mineral form. The liquid from such tanks is filtered off into a stream. Occasionally these tanks need emptying, and afterwards they take some time to become effective again as the microplants and organisms must be replaced.

A few local authorities have constructed composting plants to convert household waste and sewerage sludge into a bacteriologically harmless organic soil additive. The composting is carried out in special containers where the waste remains for five days at a temperature of around 60°C. The compost produced in this way will provide valuable humus to improve the fertility of both sand and clay soils, and remove a source of public embarrassment at the same time. (See also section on 'Organic manure', in Chapter 17.)

(E) Useful micro-organisms in industry and agriculture

(a) Cheese and butter manufacture

Milk is a very unstable food. For a long time it has been understood that the natural souring of milk produces lactic acid which protects the curd or cheese formed from further deterioration. This curd, protected from aerial contamination by a wrapping, will mature into cheese which has a sharp flavour, good keeping quality, and contains most of the valuable proteins and vitamins which are found in milk.

Cheese manufacture

Two stages in the change from milk to cheese are important. First the milk is curdled, i.e. the milk proteins are coagulated. Second the curd is ripened by the action of bacteria or fungi which produce by-products that flavour and preserve the cheese.

Stage one can be microbial, as lowering the pH by producing lactic acid automatically curdles the milk. More often curdling is caused by rennet, an extract of the enzyme rennin found in the wall of a calf's stomach.

Stage two varies according to the type of cheese to be made. Cheddar cheese is flavoured and preserved by the activity of lactic acid bacteria. Blue cheeses, like the Midland Stilton, are ripened by a fungus growing in the curd. Special strains of the correct fungus are nurtured by the cheese makers for inoculation into the curd. The French Camembert cheese is ripened by a fungal mould growing on the outside of the cheese. Swiss Gruyère cheese is ripened by propionic acid forming bacteria which also produce carbon dioxide, hence the cheese ripens with holes in it.

Butter manufacture

Cows' milk contains 3–6 per cent butterfat which can be separated from milk by agitating the liquid. Milk is a suspension of fat droplets in a water solution of proteins, vitamins, etc. Butter is a suspension of water droplets in fat.

The separation of the butterfat from the milk is effected in two stages. First the milk is skimmed, this produces cream and skimmed milk. the cream is pasteurised to kill the bacteria present, then a fresh starter culture of lactic acid bacteria is added to it. The development of these bacteria speeds the separation of the butter from the cream, and imparts flavour to the product.

(b) Organic matter breakdown in soil

Fungi and bacteria are responsible for the breakdown of the complex chemicals present in decaying animals and plants. As a result of their activities, only minerals remain in the soil ready to be reabsorbed by green plants.

The first stages of this process occur on the surface of the ground or in a well aired soil, when fungi of different kinds digest the important bonding materials holding the body of the plant together. Bacteria and insects play an important part in this early stage. Insects are very important in animal waste breakdown.

The process is continued by more fungi and bacteria. The large protein, fat, and carbohydrate molecules are further digested to release the constituent small ingredient molecules. These molecules may undergo further changes until they appear in that form which is available to plants. The activity of these micro-organisms produces much carbon dioxide to replenish the stock in the air.

The breakdown of protein in a fertile soil provides a good example of mineralisation.

1. Protein is putrefied by bacterial action to form ammonia.

2. Ammonia is oxidised to form nitrite by nitrifying bacteria called *Nitrosomonas*.

$$NH_3 \longrightarrow NO_2$$
(ammonia)　　　　(nitrite)

3. Nitrite is oxidised to nitrate by more nitrifying bacteria called *Nitrobacter*.

$$NO_2 \longrightarrow NO_3$$
(nitrite)　　　　(nitrate)

Nitrate is the form in which plants absorb their nitrogenous nutrient to make their protein, so the cycle is completed.

Winogradsky first discovered the existence of these nitrifying bacteria in 1890. Other renowned bacteriologists studied the movement of nitrogen compounds in the soil, notably Louis Pasteur—one of the greatest microbiologists of all. They discovered that certain free-living soil bacteria changed nitrogen from the air into soil nitrate available to plants. One such bacterium is *Azotobacter*, named after the French word *azote* meaning nitrogen.

Another type of bacterium living in association with the roots of legumes like clover, lucerne, or beans, was found to be responsible for turning the virtually useless nitrogen gas in soil air into a compound usable by legume roots. These micro-organisms, called *Rhizobium*, live symbiotically in the nodules formed on legume roots. Symbionts are organisms which take food from their hosts but provide something useful in return.

(c) Rumen digestion

No mammal produces an enzyme to digest cellulose the carbohydrate fibre from which grass is made, but many bacteria do. The bacteria which live symbiotically in the cow's rumen can digest cellulose with an enzyme called cellulase which they produce. Without these bacteria, ruminants (cows, sheep, and goats), cannot digest cellulose. The first stomach, or rumen, of a calf is sterile at birth but as the animal grows bacteria enter and colonise this organ as it expands in size. A calf is provided with cellulose to pick over from an early age. This helps to encourage rumen microbes to develop.

The rumen bacteria produce heat, methane gas, and acids as waste products but most important is the production of carbo-

SIMPLE NITROGEN CYCLE

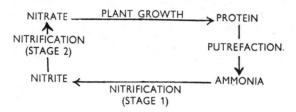

HOW ATMOSPHERIC NITROGEN SUPPLIES COMPOUNDS
FOR THE NITROGEN CYCLE

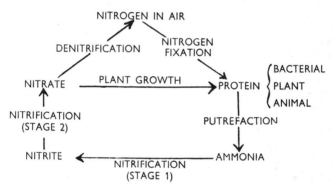

Fig. 15.6 Nitrogen cycle.

hydrate and protein for their growth. The bacteria constantly overflow down into the abomasum, or fourth stomach of the cow, where they are killed and the bacterial protein is digested. Later the bacterial carbohydrate is also digested and the products are absorbed and used by the cow. This also occurs in sheep and goats.

Rumen bacteria will digest vegetable or animal protein for their own protein synthesis, but they can take in urea which is a simple nitrogen compound made fairly easily in a chemical factor. This feeding technique used carefully and not to excess can release valuable plant proteins for feeding to non-ruminants like pigs and poultry.

16 Some important invertebrates

The animal kingdom is large and its members are varied in size and shape. When the word animal is used, however, many people automatically visualise a mammal, i.e. an animal with a backbone and hairy coat which suckles its young on milk.

The backboned group of animals itself is diverse, ranging from fish to mammals. By far the largest part of the animal kingdom, numerically, is divided amongst the invertebrate animals without backbones.

Some of these animals are useful to farmers, like the earthworm that aerates and improves the soil; while others are harmful, like the roundworms which infest sheep, cattle, pigs, and poultry. Unless a balance is maintained between the useful and harmful animals which dwell on or in the soil, then agriculture cannot progress smoothly.

Management of the farm, i.e. cropping and livestock policies, greatly affects the harmony of the small creatures living in the soil. A simple knowledge of their activities, effects, and life histories may help to avoid a gross imbalance of organisms. A representative cross section of important animals will be described in this chapter.

The animal kingdom is large and has been divided into many groups. The principles of classification have been covered in the agricultural botany section. The animal kingdom is divided into major groups called phyla, which are groups of animals with some fundamental character in common. For example, the chordate phylum includes only those animals which develop a notochord during their embryology. This notochord later forms the basis of the back-bone in most of the group, so the chordates are largely animals with back-bones.

Other fundamental similarities are used to divide the phyla into classes, orders, families, and finally into genera and species.

The animal kingdom can be segregated into those groups of animals which have back-bones, and those without which are called invertebrates. The important invertebrates are so varied that they are most usefully described under their classification groups. Then those animals with a common structure, like roundworms for example, can all be dealt with together whether they live in the soil, in plants, or in an animal's intestine. A classification of invertebrates is printed here to help the reader to see where the organisms, described later, fit into the general pattern.

(A) Classification of invertebrates

Phylum: Protozoa   (single celled animals)
                                       free living in the soil and
                                           parasites of farm livestock.
Phylum: Platyhelminthes   (flatworms)
          Class: Trematoda   (flukes—liver parasites)
          Class: Cestoda   (tapeworms—intestinal
                                           parasites)
Phylum: Nematoda   (roundworms)
                                       parasites of plants, called eel-
                                           worms; parasites of animals
Phylum: Annelida   (earthworms)
Phylum: Arthropoda   (jointed-limbed animals with
                                           hard skin)

          Class: Crustacea   (includes woodlice)
          Class: Myriapoda
            sub-class: Chilopoda   (centipedes)
            sub-class: Diplopoda   (millepedes)
          Class: Insecta
            sub-class: Apterygota   (wingless)
            order: Collembola   (soil animals)
            sub-class: Pterygota   (winged insects)
            Division I: (nymph similar to adult
            Exopterygota   insect)
            order: Orthoptera   (locusts)
            order: Thysanoptera   (thrips)
            order: Hemiptera   (plant bugs)

Division II:

Endopterygota	(larva different from adult with pupal stage in between; larvae are often soil dwellers.)
order: Lepidoptera	(butterflies and moths)
order: Coleoptera	(beetles)
order: Hymenoptera	(bees, wasps, and ants)
order: Diptera	(flies)
order: Aphaniptera	(fleas)
Class: Arachnida	
sub-class: Araneida	(spiders)
sub-class: Acarina	(ticks and mites)
Phylum: Mollusca	(slugs and snails)

(B) Protozoa: single-celled animals

Many protozoa are free living in the soil water where their activities probably pass unnoticed because of their microscopic size. They undoubtedly play an important role in maintaining a balanced population of organisms in the soil. Their food is composed of bacteria and organic matter.

While the activity of protozoa in the soil has not been clearly defined, those which are parasitic in the intestine or blood of mammals have been carefully studied.

Protozoal diseases of animals often cause chronic ailments which reduce the productivity of an animal to nothing. In young animals sickness can be more acute; coccidiosis can cause a high rate of mortality in chicken flocks.

Protozoa are often transmitted from one animal to another by other parasites of the same animal.

Examples of protozoal disease

Coccidiosis affects the intestine and caecum of the fowl. It is very dangerous to chicks, and can cause 20 per cent mortality in young flocks. Control measures include feeding sulphonamide drugs to the young bird in its food or water supply to check the disease until the bird can develop an immunity of its own.

The protozoa which cause the disease are called *Coccidia*.

Blackhead of turkeys. This disease is a killer for turkeys but other fowls seem to tolerate a high infection. The organism causing blackhead is called *Histomonas meleagris*, and the disease is spread by a parasitic caecal worm of fowls.

Where turkeys are run on fowl-sickened land they run a grave risk of contracting fatal blackhead.

Trypanosomiasis. The group of protozoa which cause sleeping sickness in man also cause diseases in cattle in the continent of Africa.

Trypanosoma brucei and *T. evansi* are the troublesome ones, particularly in Africa where a blood-sucking fly, called the tsetse fly, spreads them around.

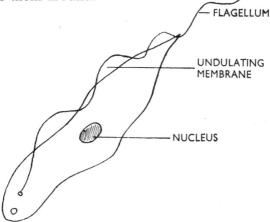

Fig. 16.1 Trypanosome, protozoal parasite actual size around 0.01 mm long.

Several attempts were made to establish herds of European cattle in Africa, but each time success was marred by trypanosomiasis. Local cattle are semi-resistant to trypanosome parasites, so more recent work has concentrated on improving the local resistant breeds.

Controlling the tsetse fly vector would eliminate the spread of this disease.

Red-water fever. This is caused by an organism called *Babesia bovis* which is spread by ticks. It is not common in the United Kingdom but can still be found lingering in marshy areas. The eradication of red-water fever is probably due to the control of ticks.

(C) Platyhelminthes: flat worms

A wide variety of flat worms are parasitic in the intestine and bile duct of domestic animals. The life cycle of some of these worms is so complicated that it is remarkable that they are still passed on. Most worms counteract their slender chances of successfully transferring eggs to another host by producing many eggs. One tape-worm can produce 150,000,000 eggs annually. When farm animals are crowded together, particularly on permanent pasture or in dirty buildings, worms can be passed on more easily.

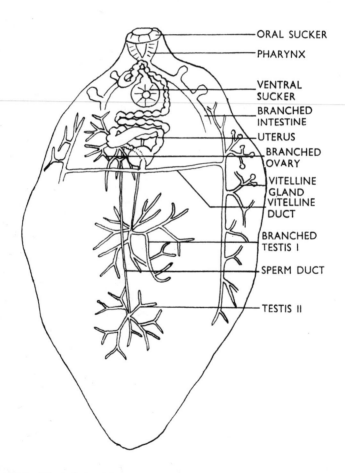

Fig. 16.2 Liver fluke (*Fasciola hepatica*) actual size 25 mm in length.

One estimate of the national loss caused by liver fluke damage each year was £1,000,000.

Animals suffering from worms eat a lot, do not thrive, and often have distended abdomens.

(a) Trematodes (liver flukes)

Adult liver flukes of sheep are about one and a half inches long and leaf shaped. They can be found in the bile duct in sheep.

Eggs from flukes pass out of infected hosts in faeces (dung). The eggs hatch to form small mobile swimming animals which must find and enter a certain type of snail which frequents damp places in pasture land. This snail is called the secondary host animal.

Inside the snail there are several changes in the structure of the fluke until, finally, an active swimming group of animals emerges called cercaria. The cercaria encyst on grass to be eaten again by primary host animals.

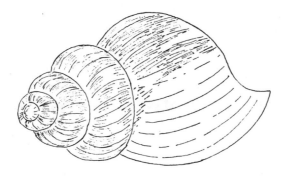

Fig. 16.3 Fluke carrying snail. Scale: snail illustrated is 8 mm in length.

To eradicate a parasite, attack it at the weakest point during its life cycle. Eliminating water snails from pasture would break the life cycle of the sheep liver fluke. Drainage of sheep pasture removes the snail host.

Poisons to eradicate the adult flukes are not always very effective. Prevention is much better than cure.

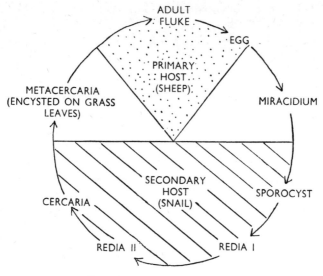

Fig. 16.4 Life cycle of liver fluke.

(b) Cestodes: tape-worms

A typical intestinal tapeworm parasite has a small knob at one end. This knob is the scolex or head. The scolex grips the wall of the intestine with rings of sharp hooks and suckers. Under a short neck, the ribbon of numerous proglottides is formed. Proglottides near to the end of the worm are ripe and filled with eggs. These proglottides drop off and pass out unnoticed in faeces. The egg is picked up and swallowed by another animal called the secondary host.

Inside the intestine of the secondary host the egg, now developed to form a small round boring animal, eats through the intestine wall to lodge in the muscles or nervous system of the secondary host. When meat from an infected secondary host is eaten by the primary host, the bladder worms, or cysticerci, from the digested muscle hook onto the intestine to become adult tapeworms.

Some tapeworms form very large bladder worms in the secondary host. *Taenia multiceps* forms multiple cysts called a coenurus with more than one scolex within. The primary host is a dog and the secondary host the sheep. The coenurus settles in the brain frequently, in sheep causing irreparable damage to the nervous system. A disease called 'gid', or staggers, is the result in sheep. Many farmers will not allow dogs on their sheep pasture because their droppings may contain eggs from *Taenia multiceps*.

The tapeworm, *Echinococcus*, is also a dog worm but humans can be the secondary host. Damage caused to the nervous system by infection with a large bladder worm cyst is dangerous.

Dogs kept as pets should be wormed regularly for both roundworms and tapeworms, and cleanliness remembered when dogs lick humans.

One animal can carry several tapeworms. Their total effect on the host is diarrhoea and wasting.

Prevention of worm spread requires a careful watch on cleanliness by a stockman.

A sow should be washed down round udder and hind quarters before farrowing. The pen where the piglets will be reared should be cleaned out and disinfected before use.

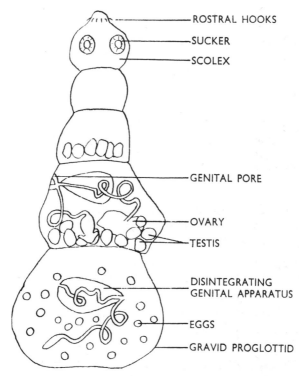

Fig. 16.5   Tapeworm (*Davainea*—poultry tapeworm) actual size 0.5–3 mm long.

Great care must be exercised when liquid manure is spread onto the grazing land. Manure from sick cattle boxes should be led away separately from the main collection channels. Manure heaps which are allowed to heat, may reduce worm egg numbers and therefore reduce sward contamination. Burying the manure under arable crops would be the best answer.

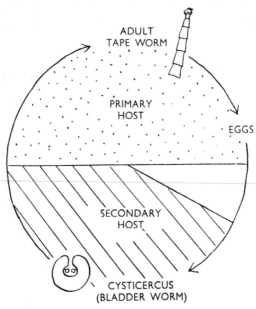

Fig. 16.6 Life cycle of tapeworm.

(D) Nematodes: round worms

Round worm parasites can be found in nearly every living animal and plant. They vary from microscopic threadworms, the eelworms of plants, to the largest round worms of animals which are over one foot long.

Nematodes can be found infecting the roots, stem, and leaves of plants; and in animals can invade the skin, blood vessels, intestine, stomach, and lungs.

(a) Animal parasites

Round worm parasites invade most of the organs in the body of a mammal. The intestine of a domestic animal and certainly that of a wild animal is almost certainly infected with round worms.

The sheep is renowned as a host for roundworms which inhabit most stages of the gut. The worms of sheep are usually fairly small although their numbers can be so great that they completely block the gut.

Cattle and pigs can carry large round worms of the *Ascaris* group. Cattle also suffer from worms in the bronchial tubes of the lungs. Such worms interfere with breathing and result in a cough developing in the host. This cough is called husk.

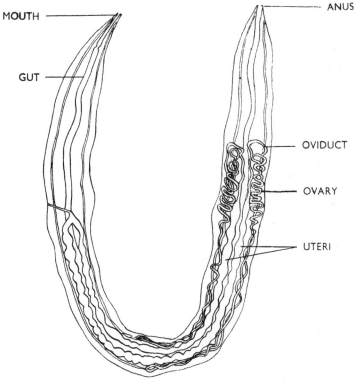

Fig. 16.7 Roundworm of pig (*Ascaris* sp.) actual size 31 cm long.

Poultry also have worms interfering with breathing, making fowls gape for air. The worm is called the 'gape worm'. Intestinal worms are common in poultry. The caecal worm, for example, is especially important as it can transmit 'blackhead', the protozoal disease, which is fatal to turkeys.

Control of worms lies largely in good stockmanship or good herd or flock management. Infected animals are best isolated to avoid spreading worms. Young susceptible stock should have the pick of clean grazing ahead of the adults. Regular rotation of grazing plots throughout the season cuts down the parasite build up.

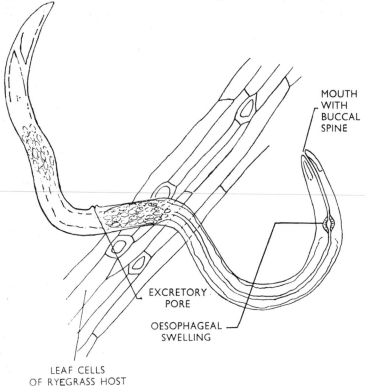

MOUTH
WITH
BUCCAL
SPINE

EXCRETORY
PORE

OESOPHAGEAL
SWELLING

LEAF CELLS
OF RYEGRASS HOST

Fig. 16.8 Stem eelworm (*Anguillulina* sp.) in leaf sheath of perennial ryegrass. Worm illustrted is 0.5 mm in length.

The husk worm eggs hatch in wet conditions, and the larvae settle on the top of moist grass. When the grass begins to dry out, the worms leave the grass. Thus, keeping young stock off moist morning grass could prevent them from contracting husk.

Chemical control at present is quite good. Drenches of phenothiazine and, more recently, thiabendazole, give fairly good clearance of roundworms.

Calves treated with dormant cultures of live husk round worms gain an artificially acquired immunity to further natural infections. This technique should always be used well before young calves are turned out to grass in the spring.

(b) Eelworms

Unlike the large animal parasites, the roundworms attacking plants are all microscopic. Parasitic plant eelworms, as they are called, have a buccal spine which can pierce cells.

Some female eelworms form cysts when ripe, which can be windblown from one field to another or even carried on lorry wheels for many miles to be dropped off on clean land.

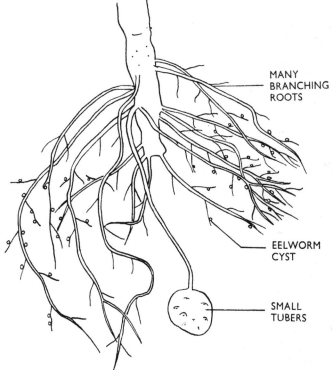

MANY BRANCHING ROOTS

EELWORM CYST

SMALL TUBERS

Fig. 16.9 Potato root infected by potato root eelworm.

Land becomes potato sick or sugar beet sick when the eelworm numbers are high.

The potato and sugar beet root eelworms are best known to

farmers because of the crop losses attributed directly to them.

The cereal root and cereal and grass stem eelworms may, however, be assuming more importance as farmers move towards longer cereal rotations.

Land which is badly affected by the potato root eelworm may have to be rested for ten years to free it from the worm cysts. A rotation of one potato crop in four years prevents any serious build up. The beet eelworm cysts are more readily starved out of the soil, but a break of two to three years is still the best control measure.

When resistant cysts form in the soil, counts can be taken and more than three cysts per 10 gm of soil is borderline for sickness in a following crop.

(E) Annelida: earthworms

There are several species of earthworm in the United Kingdom's agricultural soil, but the two most important families or genera are the grey *Allolobophora* and the red *Lumbricus*.

They live in burrows which they dig through the soil by either pushing the soil aside or swallowing it. The burrows can extend underground for several metres. The worms hide at greater depth during drought or winter cold.

At night the earthworm comes to the surface to feed, mainly on vegetable matter but also on animal remains. Thier food is digested in a long intestine by enzyme action. Calciferous glands near the front of the gut secrete calcium carbonate. A shortage of lime or calcium carbonate in the soil, associated with soils of low pH, will discourage the active earthworms which can no longer obtain calcium carbonate for their glands. If their burrows are flooded, the worms come to the surface; many are drowned in water logged soil. Well-drained land, therefore, favours earthworm activity.

Earthworms are hermaphrodite, i.e. have male and female reproductive organs in the same animal, but they exchange seminal fluid containing spermatozoa with another worm. This mating occurs at the surface of the ground usually after rain, early on a warm morning. Hence the saying that, 'the early bird catches the worm'.

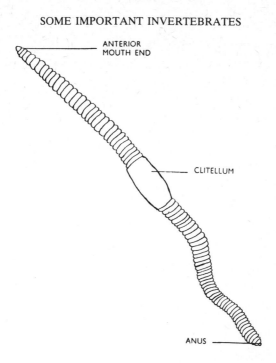

Fig. 16.10 Lumbricus terrestris, the earthworm actual size 18 cm in length.

The worm stores the sperm obtained from its mate in special sacs and releases small amounts to fertilise groups of eggs which it lays and wraps in a lemon-shaped cocoon after fertilisation. A single worm usually emerges from each cocoon although it contains several eggs.

Worms live for about five years in the soil if conditions are favourable.

The activities of earthworms

The eminent British biologist, Charles Darwin, published a work on earthworms in 1881, just before he died. Much of our knowledge of the habits of earthworms is based on his work. The agriculturally important activities are as follows:

1. Their burrows drain and therefore aerate the soil.
2. They mix organic matter with soil.
3. They speed up the mineralisation of organic matter by stimulating the activity of bacteria in their intestines.

4. They neutralise acids in soils.
5. They deposit finely divided soil on the surface of the ground in which seeds germinate easily.

Numbers of earthworms

Darwin estimated that between 60,000 and 120,000 worms per hectare was normal in U.K. soil. More recent estimates put the figure between 1,250,000 and 6,250,000 per hectare. However, probably less than half this number are active soil cultivators.

The value of earthworms

Earthworms are most valuable under grass since their cultivations aerate the soil and their castings on the surface allow seeds of perennial grasses to germinate.

Without adequate worms in a soil, grassland could deteriorate rapidly in quality.

Nitrogen fertilisers containing ammonium sulphate tend to make the soil acid and discourage earthworms because of this. Regular liming could counteract this effect.

It is certainly worthwhile doing a little to encourage these unpaid assistants of the farmer as their contribution to soil fertility is invaluable. André Voisin, French biochemist and grassland expert, drawing attention to their valuable work in grassland, alluded to them as the "Lilliputian ploughmen" of the soil.

(F) Arthropoda: jointed-limbed animals with hard skin

All the arthropods thrive in humid conditions which they seek out in the soil or under stones. Many of them require water for their larval development.

The class crustacea is largely a water group including crabs and lobsters, but the woodlice and garden slaters are members of this group. They are always found in moist conditions near the surface of the soil. They are of no great economic importance.

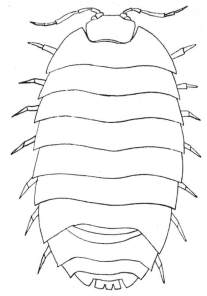

Fig. 16.11 Woodlouse, actual size 15 mm in length.

(a) Millepedes and centipedes

The class Myriapoda includes the Chilopoda (centipedes) and the Diplopoda (millepedes).

The centipedes are useful animals; carnivorous in habit, they destroy numerous small soil pests.

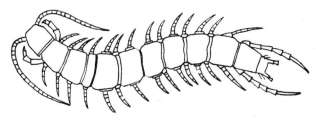

Fig. 16.12 Centipede, actual size 24 mm in length.

The millepede is a vegetarian and feeds on roots, but is not usually more than a minor soil pest.

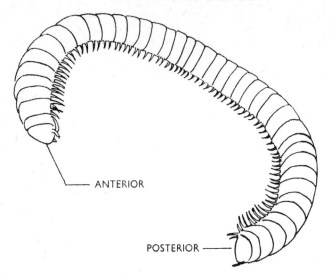

ANTERIOR

POSTERIOR

Fig. 16.13 Millepede, actual size 25 mm in length.

(b) Insects (six legs): important pests of crops and stock

The class Insecta contains many animals of economic importance in agriculture. Of the winged insects, the first major division is called the *Exopterygota*. All these insects have no cleat-cut change from the larval stage to the adult. Included here are the locusts, thrips and plant bugs.

Locusts and grasshoppers

The Orthoptera is the grasshopper order, but it includes the locust which in sub-tropical zones can cause devastation to crops when it migrates in swarms.

Thrips

The Thysanoptera is the thrips order. These small black insects, called thunder flies by countrymen because of their appearance in large numbers in humid weather, cause damage to flowers. The pea crop can be damaged by thrips, which invade the flower and damage the developing seed pod.

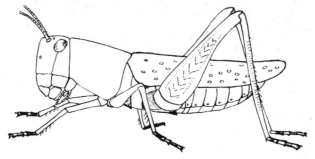

Fig. 16.14 Locust or migratory grasshopper actual size 25–30 mm in length.

Plant bugs

The order Hemiptera is an important group containing many plant pests of economic importance. These insects are small plant bugs which suck sap and, like the tsetse fly and mosquito, they are important as vectors (carriers) of disease.

The most important family is that of the aphids.

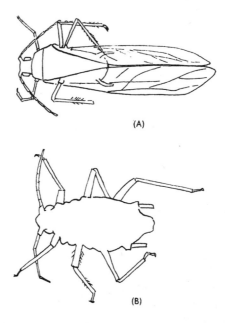

(A)

(B)

Fig. 16.15 Aphid (a) winged aphid actual size 4.5 mm in length including wings;
(b) wingless aphid actual size 1.5 mm in length.

Both the greenfly and the blackfly are aphids capable of spreading plant virus diseases. Their method of feeding by sucking sap from plant leaves weakens young plants and may cause their death in dry weather. they produce a sticky secretion called honeydew, sought after by ants and bees, which coats the plant leaf and can have a damaging effect. By far the most serious result of aphid infestation on agricultural crops is the spread of certain virus diseases by the aphid vector.

The potato crop is attacked by a green aphid called the peach and potato aphid. This insect propagates leaf roll and virus mosaic of potato.

Since potato seed tubers are produced vegetatively from the parent plant and will therefore carry any disease which was present in the sap of the parent, no seed should be saved from crops infested by aphids.

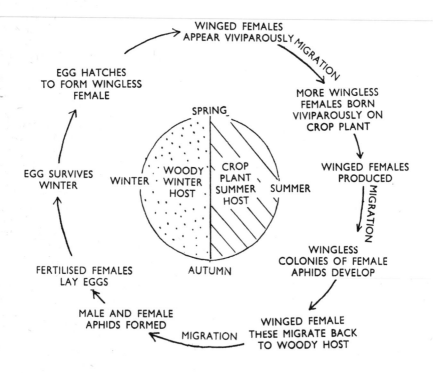

Fig. 16.16 Life cycle of aphid.

Certified potato seed which is available to the U.K. farmer has all been produced in the north of England, Scotland, or at altitudes over 150 or 170 m in southern England where aphids do not migrate successfully.

The black bean aphid attacks the flowers, young beans, and stems of beans. It also attacks sugar beet plants and spreads the virus yellows disease which causes a serious reduction in crop yield.

Systemic insecticides are used on the sugar beet crop to reduce the spread of virus yellows. A systemic insecticide renders the plant sap poisonous so that any feeding insect dies, thus preventing that insect from spreading any disease further.

Weeds in the same botanical family as beet, i.e. chenopods such as fat hen and orache, can act as alternative hosts to black aphids. Similar crop plants like spinach also harbour the pest and these reservoirs form the nuclei of major attacks on local crops.

The absence of males for most of the annual breeding cycle and the live birth, viviparity, make the aphid life history unusual. When a female is about to produce offspring her abdomen is distended with the small fully-formed aphids. Slight pressure releases the aphids in considerable number. The young are similar to the adult as there is no metamorphosis in the aphid family.

The second division of the insect class is the *Endopterygota*. The larval insect differs from the adult and the two stages are separated by pupation. Many of the larval forms in this division live in the soil.

Butterflies and moths

The order Lepidoptera includes several moths whose larval forms dwell in soil feeding on plant roots and young seedlings. These larvae are called cutworms. The larva of the yellow underwing moth is a common soil inhabitant causing damage to plant roots.

Fig. 16.17 Yellow underwing moth larvae, soil cut worm actual size 20–25 mm long.

The larva of the cabbage white butterfly causes extensive damage to the brassica crops annually by feeding on the leaves of the plants. Spraying the infested crop plants with organophosphorus based insecticide renders the plant sap poisonous to the caterpillars for a specific period and is effective in controlling an attack.

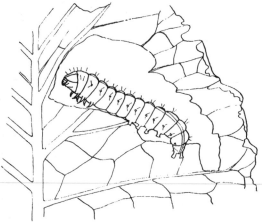

Fig. 16.18 Active cabbage white butterfly larva (caterpillar) actual size 20–30 mm long.

Beetles

The order Coleoptera is easily recognised by the hard wing cases (elytra) which cover the second pair of membranous wings underneath.

Some beetles are useful to farmers; the ladybird is carnivorous, feeding on aphids. Others have been used to bring troublesome weeds under control.

Many beetles are harmless, but some families are definitely troublesome.

The flea beetle is the black or yellow striped tiny black shiny beetle which jumps onto brassica seedlings (turnips, swedes, cabbage, kale, etc.) and causes small holes to appear in the seedling leaves. These punctures weaken and often kill seedlings. Older plants grow tougher skins which resist the attacks of the flea beetle.

The turnip fly, as the flea beetle is often called, is most active on warm June days just when swede and kale crops are coming

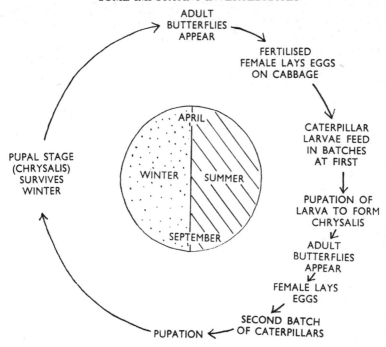

Fig. 16.19 Life cycle of cabbage white butterfly.

through the ground. The charlock plant, which is a common weed, is used as a reservoir host plant for hungry flea beetles until they can find a brassica plant to feed on.

Control of the beetle is achieved by dressing the seed with insecticidal dust before planting, then repeating the dust treatment on the young seedlings if necessary.

Weevils are beetles—easily identified by their small size, pointed snout and elbow shaped antennae. Farmers know them as leaf and seed eaters. Weevils attack and damage the seed pods of clover and oil seed rape and also feed on the leaves of clover, pea and bean plants. They also damage stored grain and the holes they leave often encourage other pests to enter. Much spoilage of stored grain begins with the activities of small insects like the grain weevil. The production of heat and moisture through insect respiration can trigger off the germination process in surrounding grain.

Grain weevils can be controlled in corn bins by fumigation.

There are several types of click beetle in the U.K. They are all small (about 1 cm in length), dark in colour, dull, and have parallel-sided elytra. If picked up they jump vigorously by flicking their abdomen backwards with a loud click—hence their name. The larvae of the click beetles are called wireworms. These larvae live in the soil for five years, feeding on plant roots and stems below ground level. When fully grown these orange, tough-skinned grubs are about 20 mm long or slightly more.

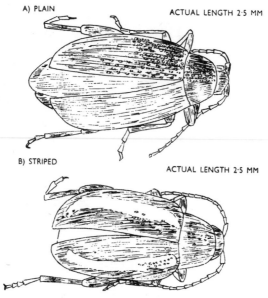

A) PLAIN   ACTUAL LENGTH 2·5 MM

B) STRIPED   ACTUAL LENGTH 2·5 MM

Fig. 16.20   Turnip flea beetle.

The beetles can be found on the ground between April and July. They seem to prefer grassland for egglaying; as a result, the damage caused by the wireworm to seedling plants is worst in the first season or two after ploughing up pasture.

All the cereals are attacked by the larva but the level of damage caused is determined by several factors such as stage of plant growth in April, pH, oxygen content of soil and availability of nutrients.

Control of wireworms was quite effective while the cereal seed was being dressed with chlorinated hydrocarbon insecticides. The indiscriminate use of such compounds has been curtailed by

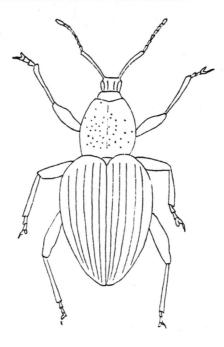

Fig. 16.21 Weevil, actual size 8 mm long.

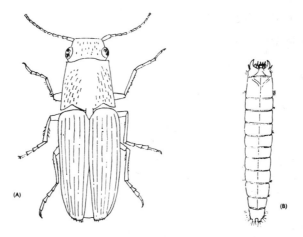

Fig. 16.22 (a) Click beetle adult, actual size 11 mm in length;
(b) Wireworm, click beetle larva, actual size 5–20 mm in length, with 1 pair of
legs on first three (thoracic) segments.

industry, as the side effects being caused to the balance of nature were causing concern among experts. Seed dressing is still the best answer to the wireworm problem, but care must be exercised in selection of insecticide.

The cockchafer or May bug, is one of the largest beetles in the U.K. Its larva, the white grub, is also very big, and since it lives in the soil for three years eating plant roots and stems it can cause damage to crops.

The adults appear in May and are noticed because of their large size and clumsiness.

The larvae become fully grown after three years of feeding in the soil. During their last year they become crop pests reaching 50 mm in length.

Chafer grubs mostly damage old grass swards but will attack sugar beet and potatoes. Ploughing and discing operations greatly reduce numbers of grubs by physical damage and exposure to predators like the seagull which follows the plough.

Bees, wasps and ants

The order Hymenoptera is best known for the sting which many of these insects carry. To the farmer, this group is very important for the successful pollination of many agricultural and horticultural seed and fruit crops.

The honeybee is the most important pollinating agent in the group. Without its help, the clover seed crop would not appear and many fruit trees would not develop any fruit. Hives of bees are usually placed in a clover crop being grown for seed or in an orchard at blossom time.

Wasp adults cause top fruit spoilage, attacking plums, apples, and pears. Control is best effected by finding their nest and gassing them.

Anthills can be a nuisance on grassland but where land is ploughed regularly, ants present no problem in the U.K.

The larva of the sawfly is damaging to the apple crop. The adult looks like a small bee, it lays eggs on the young developing fruit, and the larva feeds inside the fruit, eventually pupating and then cutting its way out. One species of sawfly attacks wheat stems, but is of no great significance.

The true flies

The two-winged flies are well known as vectors of disease among animals and are worriers of livestock.

The habits of the housefly cause bacteria to be spread from dung heaps to animal food. Milk cleanliness is more difficult to maintain when flies are about. Stock are more irritable when flies pester them; milk yields suffer as a result of the irritation.

The control of houseflies is very difficult. Regular spraying with insecticide can reduce the numbers of flies in a closed building such as a cowshed.

The warble fly and gadfly are parasitic insects. The adult gadfly sucks blood from animals and man. Its bite is painful and when they are about, usually in humid summer weather, stock become restless and may gad. Milk yields fall when cows start gadding.

The warble fly has markings similar to a wasp in colour. The eggs are laid on the hair of the flanks of cattle. The larvae hatch and burrow through the body tissues to the outside of the oesophagus. They lie dormant for a time then move through the muscles to the centre of the back in February and March.

Apart from the loss of milk and liveweight gain caused by gadding, the migration of the larvae causes pain like rheumatism which again lowers production levels.

The hide of a warble infected beast is damaged and therefore of little value for leather, and meat around the warbles is gelatinous.

Derris powder in water, rubbed on the back, poisons the larvae.

Sheep strike, or subcutaneous myiasis, is caused by greenbottle fly maggots eating the skin. The wool drops and the raw wound is infested by other fly maggots. Humid weather suits the fly larvae so this disease is common in hill flocks especially where the fleece becomes soiled around the breech area.

The larvae are only on the skin for two days; after this they are fully fed and drop to the ground where they pupate. The severe setback suffered by the sheep, especially if it is a young animal and possibly carrying intestinal parasites in large numbers, could prove fatal.

In Africa, the tsetse fly spreads the trypanosome protozoal parasite. This destroys red blood corpuscles and the animal suffers from anaemia. The fly bites an infected beast and carries the parasite from the blood of one animal and injects it into a healthy animal. Attempts to introduce European cattle to a tsetse fly region have proved unsuccessful because of the trypanosome parasite. (See under 'Protozoa' for description of trypanosome.)

The female crane fly, or daddy longlegs, lays its eggs on moist soil under grass in August and September. After a fortnight the eggs hatch to produce a greyish larva without legs, called a leatherjacket. These larvae feed on roots just below the soil surface throughout the autumn, winter, and spring, before pupating in the summer and producing adults again.

Leatherjackets can be very damaging to the cereal crops.

Treating the soil with bran bait mixed with organo-phosphorus insecticide can destroy many larvae when the infestation is heavy.

The larvae of the wheat bulb fly attack autumn sown wheat.

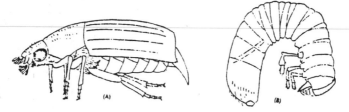

Fig. 16.23 (a) Cockchafer (May bug), scale □ 1 mm; (b) Cockchafer larva (white grub in soil), scale □ 1 mm.

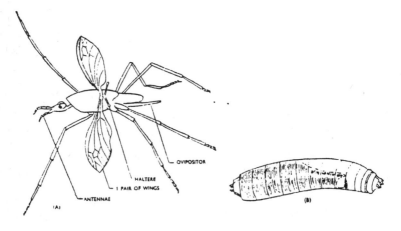

Fig. 16.24 (a) Crane fly adult, actual size head and body 20–25 mm long; (b) Leatherjacket, crane fly larva magnified, actual length 28 mm.

Egg laying occurs in August, but the larvae do not hatch from the egg on the soil until January or February. The larvae burrow into the wheat tillers destroying the growing point of the stem. They will move from one tiller to another until fully fed, in April or May, when they drop off the plants and pupate. The adults appear in June.

Severe attacks can wipe out areas of winter wheat. Barley and rye are also attacked, but oats are immune.

Severe attacks of gout fly appear when a crop of barley is sown late in the spring. The flies appear in May and June, laying their eggs on the leaves near to the centre of the barley plant. The larvae burrow down to the first stem joint to feed. When fully fed after a month they drop out or pupate where they are. A second brood can appear in the same year, but the eggs are laid on couch grass and the larvae pupate until spring.

The effect on the barley is to cause the ear to be damaged or not to appear at all. The stem swells up where the larva is feeding and excessive tillering occurs.

Frit fly maggots can severely damage late sown crops of oats. The flies appear in May and lay eggs on oat plants. The larvae burrow into the shoots causing the plant to branch out more tillers. The maggots, fully grown by June, pupate and a second generation of flies attack the ear and the larvae feed on the grain. These maggots pupate and produce adults which lay eggs on perennial ryegrass. The eggs hatch and survive the winter in the larval stage in the ryegrass.

Winter oats are able to withstand attacks as they are better established by the time the flies arrive.

Fleas

The order Aphaniptera are all parasites feeding on the blood of birds or mammals. They live on the skin of these two warm-blooded groups of animal (fig. 16.26).

They have no wings and are flattened and tough. Their legs are powerful and they hop to avoid being removed.

Eggs called nits are fixed to hair base on laying.

Fleas can be destroyed by spraying the hair of an infested animal with organo-phosphorus insecticide.

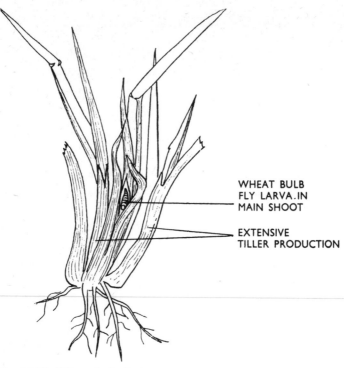

WHEAT BULB
FLY LARVA.IN
MAIN SHOOT

EXTENSIVE
TILLER PRODUCTION

Fig. 16.25 Wheat bulb fly larva on an autumn sown wheat plant.

An animal suffering from a heavy infestation looks out of condition and may be anaemic. In poultry, the comb loses its colour.

(c) Ticks, mites, and spiders

The class Arachnida are arthropods with eight legs. The spiders, although of no great importance, are a good example.

The sub-class Acarina, the ticks and mites, are pests of livestock and plants.

Hard ticks

Ixodes ricinus is the common tick in this country. It is associated mainly with sheep, but can climb onto cattle as well.

The tick has three separate feeding stages in its life; between

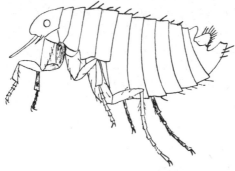

Fig. 16.26   Flea, actual size 4 mm long.

these it drops off the host and may lie on the ground for six months. The feeding periods last about three weeks at a time.

Because of their habit of climbing onto different hosts to feed on their blood through the skin, they are responsible for the transmission of several diseases. Louping ill, tickborne fever, and red-water fever are all transmitted by the tick.

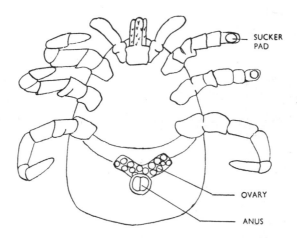

Fig. 16.27   Sheep tick larva (ventral view), actual size 6 mm long, nymph and adult tick have 8 walking legs.

Mites

These small tough arachnids cause mange on farm livestock. Treatment with organo-phosphorus insecticide helps to control them. Poultry suffer from red mite which attacks them as they perch at night. During the day the mites hide in crevices of the perch. Paraffin treatment of the perch and creosote on the walls of a poultry shed help to cut down mite numbers.

Mites are common pests of fruit trees and bushes. The big bud mite of blackcurrants destroys the developing flowers and spreads a plant virus disease, called reversion, which ruins the currant bush.

Mites are very difficult to kill. The organo-phosphorus sprays are used with most effect, but these are the most dangerous of the sprays used in horticulture from the operator's point of view.

Mites which infest grain stores can be removed by fumigation with ethylene dichloride. This job is more often carried out by a contractor.

Sheep scab, which is a notifiable disease, is caused by mites which bite the skin of a sheep and feed on lymph. The wounds form scabs and the irritation caused to the animal leads to emaciation. The condition is highly contagious so compulsory autumn dipping in an approved acaricide (poison specifically for the Acarina, i.e. ticks and mites) is carried out to suppress the mites which cause the disease.

(G) Mollusca: slugs and snails

This group of animals is widespread through the world and appears far back in geological time. Close aquatic relatives are the squid and octopus, and some snails still live in water. One of these small water snails is host to the sheep liver fluke parasite between its passage from one sheep to another.

Those slugs and snails which live on land are mostly vegetarians, and therefore are crop pests. They are only found in moist conditions, so present their greatest challenge to the farmer on heavy clay soil. Weed coverage favours their movement about the soil.

Young seedlings attacked by slugs often die as they lose turgidity when the stem is damaged just above the soil.

Potato tubers are attacked on heavy land and the quality of the crop is reduced as a result.

The eggs of the slug are about 2 mm in diameter, translucent and colourless. They are covered by a sticky secretion and are usually found in groups.

Slug control can be very effective if micro-granules containing methiocarb are scattered on the surface of the soil when the crop plant is most vulnerable to slug attack. The treatment may have to be repeated if the chemical disappears while the crop is still vulnerable.

Section VI

Chemicals in crop production

Introduction

To achieve the best results from fertilisers and spray chemicals, it is clearly necessary for the farmer to understand how to use them, safely, effectively and economically.

A knowledge of the principles governing their use, will enable a farmer to adapt his methods to include new techniques as they arise.

17 Fertilisers

(A) Principles of manuring

(a) Plant nutrients and deficiency symptoms

It is difficult to tell the exact effect of plant nutrients until they are in short supply. This provides one method for discovering which nutrients are required by certain crops. By limiting the supply of one nutrient chemical at a time, abnormal growth is induced in a plant. If the plant grows normally when the nutrient is replaced, then it is essential for its growth. It should be noted that some other factor could be responsible for the abnormal growth. A control plant, given an adequate supply of the nutrient, should be grown near to the experimental plant so that a comparison is possible and a watch can be maintained for any environmental effects on both plants.

Another way of determining the nutrient requirement of a crop is by soil analysis. Soil is analysed before sowing a crop and then again after harvest. The minerals removed from the soil can then be compared with those found to be present in the plants. Unfortunately it is thought that many plants indulge in luxury consumption of some minerals in the soil, over and above their growth requirements. This appears to be true in the case of potassium.

Major plant nutrients

Between six and eight plant nutrients are required in regular amounts by most crops with a further five or six needed in small quantities only.

Nitrogen has the most dramatic effect on the leaf growth, especially in grass and cereal plants. It is used to make protein in plants. Protoplasm made from protein forms about half of a plant's weight. Older leaves have their nitrogen compounds removed by the plant to supply the young growing leaves. Thus symptoms of nitrogen shortage occur first in the older leaves.

Phosphorus is essential in the nucleus of every cell so growth cannot continue in its complete absence. It is particularly associated with the development of a strong root system and with floral development.

Calcium is used to make cell walls and to neutralise waste acids. It is not readily moved about the plant so deficiency symptoms appear in the young leaves first.

Magnesium forms the centre of the chlorophyll molecule. It is not commonly deficient at modest cropping levels in the United Kingdom, but intensive cropping or excessive potash use can reveal shortages.

Potassium helps the plant to move carbohydrate and build protein for protoplasm.

Sulphur is needed for protein formation. Certain essential amino acids contain sulphur.

Minor plant nutrients

A number of minerals are absorbed by plants in minute quantities. They are called trace elements for this reason, but without them the plant cannot grow properly. They are possibly required to replenish stocks of tired catalysts which are controlling the speed of reactions in the plant.

Catalysts are rather like the oil in a four-stroke engine, very necessary for efficient running but not used up to any great extent. Occasionally, like the oil in an engine, catalysts become tired and have to be replaced.

Manganese can be deficient in soil with a high pH, i.e. soil with a high lime content. Overliming soil could lead to this kind of trouble.

Boron seems to be important in the absorption and utilisation of major nutrients. Again overliming can lead to a deficiency of this nutrient.

Copper and zinc have been shown to be necessary for fruit trees and deficiencies may cause troubles in livestock.

Molybdenum is important to legumes for correct nodule formation. It seems to be required by the rhizobium bacteria. A high pH in the soil causes plants to absorb excess molybdenum. Cattle grazing grass containing excess molybdenum can be partially poisoned. The disease is molybdenumosis and the pasture causing it is described as teart.

Sodium and iron are probably really major nutrients as they are absorbed in a greater concentration than is usual for minor nutrients.

Iron is needed to make chlorophyll, although it is not present in the completed molecule. A high soil pH reduces the availability of iron.

Sodium is necessary for sugar beet and mangolds, but in other plants it may only substitute for potassium to some extent.

Deficiency symptoms and diseases

Inevitably, cutting off the supply of one vital nutrient affects most of the processes occurring in a plant since they are all closely connected. It is difficult to diagnose mineral deficiencies from abnormal plant growth since similar symptoms develop from different causes.

Stunted plants might indicate a nitrogen deficiency if no nitrogen had been added to the soil during early crop growth. Red-coloured leaves could be caused by a phosphate shortage.

Yellow leaves can be caused by too many factors to guess at anything. The yellowing of leaves should stimulate a closer examination of the crop which might reveal the possible cause. A certain type of leaf yellowing in sugar beet is caused by manganese deficiency. Experience will show whether applications of manganese sulphate as a dilute foliar spray corrects the symptom. A shortage of boron causes 'heart rot' in the roots of sugar beet and mangolds and 'brown heart' in swedes. Again experience will show whether light applications of borax can correct this condition.

(b) Nutrients in the soil

A steady supply of nitrates is available to plants during the spring and summer as the soil bacteria break down animal and vegetable protein in soil organic matter. Nitrates are very soluble so they are easily leached in high rainfall areas during the winter months.

The natural supply of phosphate is poor but, as most phosphate is slow to dissolve, an accumulation of insoluble phosphates has occurred on most agricultural land where phosphate has been applied in recent years.

The clay minerals in the soil provide a continuous but small supply of potassium and probably also magnesium. There would be very little at all where the soil is predominantly sand.

Calcium is present in liming materials, so as a nutrient it is seldom in short supply.

Fortunately most soils contain a sufficient quantity of most trace elements, but often a high pH or lack of air in the soil or unbalanced fertiliser usage prevents the plant from obtaining them.

(c) Planned manuring

Excess use of fertiliser can be wasteful as much of it may be leached from the soil. Insufficient fertiliser can limit crop yield and reduce the potential profit margin on a crop.

A planned manuring policy is advisable, linked with the results of soil analysis and local weather conditions.

Soil analysis

The results show the lime, phosphate and potash status from which appropriate dressings of fertiliser can be recommended. Soil analyses can be carried out by Ministry of Agriculture chemists or by private firms who offer this service to farmers.

By using the results carefully and critically, farmers may reduce their expenditure on fertilisers while still obtaining full yields and without lowering the fertility of their soils.

Conservative interpretation of results, particularly with regard to phosphate, is advisable e.g. 'low' or 'very low' phosphate deserves a full dressing while 'satisfactory' could have half the usual dressing to be on the safe side.

Seasonal effect on fertiliser use

A wet winter causes nitrates to be leached from the soil; additional dressings would therefore be advantageous to crops in the spring. Dry winters remove less nitrate.

(d) Units of plant food

Fertiliser recommendations can be given in kg of pure plant food nutrient per hectare. Each bag of fertiliser contains a fraction of pure nutrient as stated by a percentage nutrient on the bag.

If a 50-kg bag of nitrogenous fertiliser is labelled 33 per cent N, it contains

$$\frac{50 \times 33}{100} \text{kg}$$

or 16.5 kg of nitrogen food per bag.

Price comparisons can be made between fertilisers supplying similar nutrients on a basis of the cost per kg of pure plant food.

(e) pH and plant growth (use of liming materials)

Soil reaction, that is whether acid, alkaline or neutral, is measured on the pH scale. 1–6 is acid, 7 is neutral, and 8–14 is alkaline.

Indications of pH can be obtained by using dyes which are sensitive to pH and change colour as the pH changes from acid to alkaline. The importance of any pH test lies in sampling the field correctly, more than in the accuracy of individual results.

Plants will not tolerate a wide range of pH. Most seem to thrive in a neutral to slightly acid soil. Potatoes and oats are crops which tolerate more acidity than most, but below pH 5 they too begin to suffer. Above pH 7.5 to 8 most plants suffer setbacks in growth. This may be caused by a shortage of soluble phosphate and trace elements at a high pH.

Earthworms, vital for aerating the soil, migrate away if the soil pH drops much below neutral. The last to leave are the short red worms. The first to go are the larger species which prove the most useful in soil.

Bacteria cannot operate in a very acid soil, so humus formation is stopped and a peat develops. Legumes must have a neutral soil if their nitrogen fixing bacteria are to work properly.

(f) Liming materials

Ground limestone (calcium carbonate) is only slightly soluble in water containing carbonic acid, and must be finely ground to be useful. Two and a half tonne of ground limestone (per hectare) raises the pH by 0.75 of a unit. This will take a long time so, to compensate for this, 50 per cent more is added to speed up availability.

This material is readily available and is easily applied by contractors.

Hydrated lime (calcium hydroxide) is more active than calcium carbonate in the soil; it is used in horticulture for balancing compost pH. About three-quarters of a tonne will be equal in effect to one tonne of calcium carbonate.

Waste lime can be obtained cheaply from some industrial sources, e.g. from a sugar beet factory. It presents problems in spreading which might be overcome if it is cheap enough.

(B) Nitrogen: nitrogenous fertilisers

In the United Kingdom climate, natural nitrogen compounds are insufficient to allow plants to reach their growth potential so fertilisers are used.

The atmosphere consists of 80 per cent nitrogen gas, but plants cannot use this supply with the exception of leguminous plants.

The plant absorbs its nitrogen as nitrate through its roots. The most readily available form of fertiliser will therefore be a nitrate containing compound.

Nitrates are salts of nitric acid; they are fairly stable in the soil but very soluble and therefore leached rapidly. Nitrate forms an anion (NO_3^-) in solution so cannot be held by the clay particles.

Ammonium compounds are oxidised to nitrates in a fertile soil by the activity of nitrifying bacteria. This process may take a week or more in the soil, and longer in the winter. Ammonium forms a cation in the soil solution (NH_4^+) so it is held by clay particles against leaching.

Fertilisers supplying nitrogen

Sulphate of ammonia $(NH_4)_2SO_4$ 20 per cent N

This consists of very soluble crystals which are not very hygroscopic (absorbing water), therefore it flows well in the distributor and stores fairly well.

Anhydrous ammonia gas   82 per cent N

Liquefied ammonia can now be delivered to farms and applied directly to soil by injecting it to a depth of 150 mm, where it dissolves in the soil solution. Fertiliser applied in this way can last quite a long time.

Ammonium nitrate NH_4NO_3 35 per cent N

This is hygroscopic and inflammable. Pelleting and coating with a water repellent layer improves storage. Various grades and mixtures of ammonium nitrate fertilisers are now available, varying in strength from 23 per cent to 35 per cent.

Urea $CO(NH_2)_2$ 46.7 per cent N

This is organic nitrogen, but it rapidly decomposes in the soil to form ammonia. It requires bacterial action to change it to nitrate.

Recovery of nitrogen fertilisers

Arable crops lose more nitrate through leaching than do grass crops. The grass roots spread through the top 230 mm of soil so prove to be efficient absorbers. Arable crops often lose their fertiliser before they become established. Winter wheat usually has its nitrogen supplied in the early spring when it is established and can absorb the fertiliser more efficiently.

(C) Phosphorus

United Kingdom soils have only small reserves of phosphate naturally, but agricultural land usually has some present.

Phosphorus is far too active an element to occur in nature. The chief oxide of phosphorus, called phosphorus pentoxide P_2O_5, is formed when phosphorus burns in air. It is an acid gas which dissolves in water to form orthophosphoric acid (H_3PO_4).

$$2P + 5\,O \longrightarrow P_2O_5$$

phosphorus   phosphorus pentoxide

$$P_2O_5 + 3\,H_2O \longrightarrow 2\,H_3PO_4$$

orthophosphoric acid

It is chiefly the calcium salts of orthophosphoric acid which are used as fertilizers.

Calcium salts of orthophosphoric acid

$CaH_4(PO_4)_2$	Mono-calcium phosphate
$Ca_2H_2(PO_4)_2$	Di-calcium phosphate
$Ca_3(PO_4)_2$	Tri-calcium phosphate

Mono-calcium phosphate is water soluble; the other two are insoluble and only dissolve slowly in dilute acid.

Mono-calcium phosphate is therefore the only calcium compound which will produce phosphate ions in the soil water. The plant is only capable of absorbing phosphate ions.

$$Ca^{++} \longleftarrow CaH_4(PO_4)_2 \longrightarrow 2\,H_2PO_4^-$$

calcium   mono-calcium phosphate   dihydrogen phosphate anion

$$\downarrow$$

available to plants

(a) Calcium phosphate in soil

In soils with a high pH, soluble mono-calcium phosphate is rapidly changed to insoluble tri-calcium phosphate. A low pH favours the slow breakdown of insoluble tri-calcium phosphate into the soluble form.

High pH (lime)

Insoluble phosphate $\rightleftharpoons$ Soluble phosphate

Low pH (acid)

or: mono-calcium
                                                  phosphate

$$Ca_3(PO_4)_2 \xleftarrow[\quad]{\text{High pH} \atop H^+ + H_2O} Ca_2H_2(PO_4)_2 \xleftarrow{H^+ + H_2O} CaH(PO_4)_2$$

tri-calcium   di-calcium phosphate
phosphate $\xrightarrow[\text{Low pH}]{\qquad\qquad\qquad\qquad}$ Ca^{++} 2 H_2 PO_4
                                                  into plants

Phosphate fixation

Apart from becoming insoluble in soil, much of the phosphate forms compounds with iron and aluminium. In this form it is virtually lost for plant growth.

Reduction of fixation may be achieved by applying farmyard manure liberally to the soil.

The acids formed by the breakdown of this organic matter prevent iron and aluminium phosphates forming so readily.

(b) Phosphatic fertilisers

Mineral phosphate, bone meal and basic slag   7–18 per cent P_2O_5

All these contain tri-calcium phosphate, $Ca_3(PO_4)_2$, which dissolves slowly in dilute acid if the material is finely ground.

Basic slag is a waste product from the conversion of iron to steel. Bessemer converters produce the best agricultural slag; open hearth furnaces produce poor slag.

Basic slag with potassium added is really a compound fertiliser. It is very useful for clover, a leguminous crop.

Super-phosphate   19–21 per cent P_2O_5

All the phosphate is water soluble mono-calcium phosphate.

Lawes' process for dissolving bones and mineral phosphate is as follows:

mono-calcium
phosphate   gypsum

$$Ca_3(PO_4)_2 + 2\ H_2SO_4 \longrightarrow CaH_4(PO_4)_2 + 2\ CaSO_4$$

tri-calcium   sulphuric
phosphate   acid   super-phosphate

Triple super-phosphate   47 per cent P_2O_5

Nothing but mono-calcium phosphate is present. This is a useful compound as it is concentrated and completely soluble.

Manufacture of triple super-phosphate:

$$Ca_3(PO_4)_2 + 4 H_3PO_4 \longrightarrow 3 CaH_4PO_4$$

| tri-calcium phosphate | orthophosphoric acid | mono-calcium phosphate |

Ammonium phosphate

This is used in compound fertilisers as it contains ammonia as well as phosphate. All the phosphate is soluble in water. Two strengths are available to compounders:

1. Mono-ammonium phosphate $NH_4H_2PO_4$
   12 per cent N, 62 per cent P_2O_5
2. Di-ammonium phosphate $(NH_4)_2HPO_4^-$
   21 per cent N, 54 per cent P_2O_5

Recovery of phosphate fertilisers

Young seedlings need phosphate for root formation, so super phosphate is necessary to supply soluble phosphate. That soluble fertiliser which the plant cannot absorb quickly is soon converted into insoluble phosphate. Excessive doses of soluble phosphate can be wasteful and expensive. Slightly acid soil gives better phosphate recovery but, at best, it is slow.

(D) Potassium

(a) Natural supplies

Heavy clay soils may have a natural supply of potassium compounds, but sand soils provide nothing for the plant.

The plant absorbs potassium as a cation K^+. It can be held on clay particles so, although many of its compounds are extremely soluble in water, it is not leached readily from loam or clay soil.

Potassium is a metal when pure but, because it is very reactive, it never occurs in nature in a pure state.

Some compounds of potassium which could well be used as fertiliser are sought after by industry for other purposes. For example potassium nitrate KNO_3 is used to make gunpowder; this is a more profitable market than using it as a fertiliser.

(b) Potassic fertilisers

The potassium analysis is quoted in terms of the oxide K_2O.

Muriate of potash   60 per cent K_2O

This contains potassium chloride KCl. The pink crystals are very soluble and hygroscopic.

Sulphate of potash K_2SO_4 48–50 per cent K_2O

This is slightly less soluble in water than muriate of potash. The high chloride content of muriate of potash makes sulphate of potash more acceptable to some horticulturalists. The Dutch recorded a better potato tuber quality when sulphate of potash was used. Potash fertilisers are now available to farmers, mostly in the form of compounds.

Recovery of potassium fertilisers

Potassium is subject to leaching on sandy soil, but not as much on loam or clay. Many crops seem to absorb more potassium than they need for growth. This is not wasteful if the crop is fed to livestock in the field; but if a crop is carted off the field, then potassium will be lost.

(E) Fertiliser usage

(a) The interaction of N, P, and K

The interaction of these nutrients in a plant means that without adequate supplies of all three, any one alone will not be effective.

For example, high levels of nitrogen fertilising increase the crop requirement for potassium and phosphate. Without these two the plant cannot fully utilise the nitrogen.

(b) Compound fertilisers

These fertilisers have the advantage that each granule contains all the nutrients to be supplied to the crop in the correct ratio desired.

Compared with the price of soluble 'straight' nutrient fertilisers, compounds are very reasonable. Only one journey is necessary to apply the three nutrients to the soil.

The concentration of nutrients in compounds has increased over the past ten years. Modern compound fertilisers are concentrated to save labour in carting unnecessary bulk. This means they must be applied carefully to achieve economical results.

(c) Timing and placement of application

Potassium and phosphate must be placed near to the seed to be

1. FERTILISER BROADCAST ON RIDGES

2. POTATO TUBERS PLANTED

3. RIDGES SPLIT BACK

Fig. 17.1 Potato fertiliser broadcast on ridges before planting.

effective. Nitrogen is leached readily so is best applied when the crop is actively growing. Spring applications are effective on grassland and winter-sown cereals. Spring-sown cereals gain little advantage from having the nitrogen delayed until after sowing, so it is usually all placed in the seed bed.

Fertiliser placement

Nutrients should be placed in the soil as near as possible to the root system of a plant, without causing damage through chemical scorching.

A combine drill mixes fertiliser and seed in the soil. The young rootlet can be damaged by this contact with fertiliser so, for sensitive seedlings, side placement of fertiliser in bands near to the root but not on it is sometimes preferred.

Fertiliser applied broadcast on opened ridges, prior to potato planting, causes the nutrient to be accumulated round the tubers when the ridges are split back again after sowing.

Extra yields of tubers are attributed to this placement of the fertiliser near to the tuber.

(F) Organic manure

Plant and animal remains decay in the soil to release compounds of nitrogen, phosphate, and potassium, which can be absorbed by plants.

The nitrogen and phosphate present are not water soluble so cannot be quickly leached from the soil. As the manure decays in the ground, plant foods are released slowly at a rate which is more than matched by the uptake of the crop.

Organic fertilisers contain little or no soluble salt so they can be applied at heavy rates without risk of damaging crop roots; this damage can occur when heavy doses of inorganic fertiliser are used.

(a) Sources

The chief sources of agricultural organic manure are:
　　1. Ploughed in plant debris, stubble, etc.

2. Ploughed in pasture and ley.
3. Green manure crops, particularly legumes.
4. Farmyard manure.
5. Industrial organic waste.

Ploughed in plant debris, stubble and straw

This breaks down in the soil to supply minerals and increase the valuable humus content of the soil. The available nitrogen for plant growth is locked up temporarily by the bacteria responsible for the straw breakdown, so slightly increased nitrogen dressings can be considered for a following crop.

Straw can be composted before applying it to the soil, but the extra expense of doing this is not justified by the degree of improvement in results.

Ploughed in grass and other green manure crops

Stem and leaf of grass and clovers decay to provide nitrogen compounds for growth and humus to improve soil structure.

Slowly decomposing materials containing more stem than leaf add humus to the soil, but little nitrogen. Rapidly decomposing leafy material supplies a lot of nitrogen, but very little humus to the soil.

Farmyard manure

The quality of manure depends on the plant foods which it contains. Quality is therefore determined by the type of food given to stock and their utilisation of it. Fattening stock waste a larger percentage of their food than young animals or dairy stock, so their manure is richest in plant foods. As the straw content of manure rises, so the nutrients become more diluted.

Loss of nutrients from farmyard manure can be reduced by turning it as little as possible and ploughing it in soon after spreading. Covering the manure heap also reduces leaching. Phosphate is not lost from manure but potassium can be washed out by rain, and nitrogen is often lost as ammonia gas.

The valuable effects of farmyard manure in the soil are as follows:

1. The supply of partly decomposed organic matter increases the humus content of the soil. Humus improves water-holding and mineral-retention in sandy soil, and improves flocculation and drainage in clay soil. The organic matter also supplies food for the living soil population which is responsible to a great extent for maintaining fertility.
2. The breakdown of organic matter releases organic acids which lower the pH of the soil. The lower pH renders phosphate more soluble and some of the trace elements more available.
3. The supply of some plant nutrients.
   Nutrient contents of one tonne:

	N_2	P_2O_5	K_2O
Farmyard manure	5 kg	3 kg	5 kg
Poultry manure	13 kg	10 kg	5 kg

These nutrients are not immediately available; only one-half to three-quarters can appear during the first growing season.

Industrial organic waste

Factory waste like dried blood, shoddy (wool waste), meat and fish meal can be useful sources of nitrogen and sometimes phosphate when they decay in the soil. They must be fairly cheap to the farmer to justify their use. They form a valuable source of organic matter to horticulturalists who may not have a regular supply of farmyard manure.

Liquid fertilisers and slurry disposal

The application to the soil of both organic and inorganic nutrients in water has become a widespread practice. Although still in its infancy in the United Kingdom, early results seem to point to a promising future for some of these developments.

Liquid mineral fertiliser can be applied either to the soil around a plant or, in some cases, to the plant's leaves. The foliar spray (leaf application) has proved useful for correcting trace element deficiency symptoms which appear in a crop.

Soil application of concentrated fertiliser solutions allows the crop to become established in its seed bed before the fertiliser is added. In some cases tines have been fitted to the tool bar behind a tractor and fertiliser piped directly into the soil next to the growing plant roots. This technique is essential if liquid ammonia is used as a fertiliser.

Concentrated liquid fertiliser usually burns plant leaves, so flexible pipes should be fitted to the spray nozzles in order to carry the liquid directly to the soil surface.

Under the Rivers Pollution Act of 1960, it became an offence to discharge liquid manure into a stream. The introduction of a liquid slurry tank poses some emptying problems immediately. The most profitable way of disposing of this material is to apply it to the land. Tankers, pumps, and high pressure irrigation rain guns have been designed to apply slurry to the fields.

Winter dressing of pasture land with slurry is fairly safe for later grazing, since the tainting effect will be reduced by heavy winter rain.

During the summer grazing period, slurry can be put on pasture just after it has been grazed. The young leaves then develop after the manure has been applied and thus avoid contamination. A three week rest period followed by intensive grazing gives best utilisation of treated pasture. 20,000 litres of slurry contain approximately:

$$26 \text{ kg of N and } 26 \text{ kg of } K_2O$$

(G) Residual value of fertiliser

An outgoing tenant is paid compensation for the unused residual effects of fertilisers which he can prove that he has applied.

Nitrogen has no residual effect because it is leached readily. Phosphate is paid for at the rate of two-thirds for the first year after application, one-third after two years, and one-sixth after three. Basic slag or insoluble phosphate is worth half as much as the soluble phosphate. Potassium disappears after two crops. Lime payment is reduced by one-eighth for each crop taken.

These are only the legal time limits set on nutrients; these materials may, in fact, exist in the soil for much longer in certain soil conditions.

18 Chemical crop protection

Safe use of chemicals

Modern husbandry uses chemicals widely to reduce the number of cultivations necessary in growing a crop and also to increase the yield by removing pests, diseases and competitive weeds.

They can be used safely if they are properly handled, mixed and applied in accordance with the maker's instructions wearing protective clothing if required.

Always spend time on reading the maker's precautions and instructions before opening any container of chemical. Avoid keeping old stocks of chemical and remember to wash out empty containers into the spray tank, then perforate the cans before disposing of them in an approved dump.

The risks from using chemicals should be considered in the following order:

1. the danger to the operator;
2. the danger to the general public;
3. the danger to the environment;
4. the danger to the crop.

The weed, disease or pest is the only thing that should be damaged by the chemical if it is properly applied.

(A) Herbicides

Herbicides, weedkillers, attack plants through the aerial or subterranean parts.

Aerial application:

1. Contact weedkillers move into the plant by diffusion only, and kill the parts on which they fall.

2. Translocated weedkillers are moved about the plant in the food stream.

Soil surface application:
Chemicals remain in the top few millimetres of soil for a period of time during which they inhibit the establishment of seedling weeds.

Incorporation in the soil:
Such chemicals must be persistent and remain active in the soil for some time. The residual effect can limit the farmer's choice of following crops.

Selectivity of herbicides

Non-selective weedkillers will remove all plant growth from the soil. The more persistent chemicals will keep the ground free from plants for long periods of time. On gravel paths this could be desirable.

Other chemicals remove all surface growth through a desiccating effect but their effect is short-lived, only a few hours in some cases. These chemicals are widely used to burn off pasture for reseeding or to remove potato haulm before lifting the tubers.

Herbicides for use with crop plants must be highly selective. Factors controlling selectivity are:
1. Dilution of herbicide
2. Size of droplet (can be increased by lowering surface tension of liquid, with soap etc.)
3. Surface of plant
4. Angle of plant leaves
5. Time of spraying in relation to crop growth
6. Reaction of plant enzymes to herbicide, e.g. M.C.P.B. is converted to the weedkiller M.C.P.A. by most weeds. Clovers do not convert M.C.P.B., which is harmless, so they are unaffected by it.

(B) Insecticides and acaricides

Insecticides which are coated on the plant surface kill insects with which they come into contact, but are short-lived if the rain washes the leaf.

Insecticides which are absorbed by the plant are called systemic, since they enter the food channels of the plant and make its sap poisonous to sucking insects.

Soil insecticides are required to persist for a certain time in the soil until the insect makes contact with them. Unfortunately some of the organo-chlorine insecticides have proved to be so persistent that the levels are building up in some soils.

Field mice and pigeons eat a lot of grain which has been planted and dressed with insecticide, so they often pick up large doses of the insecticide. Foxes which eat these pigeons, and falcons which clear away small rodents, have died through eating several heavily contaminated animals. The use of these persistent insecticides is now controlled more closely to prevent unnecessary losses of wild life.

Insecticides and acaricides, as a general rule, are more dangerous to man than are the herbicides. Several are classified as dangerous poisons and the correct protective clothing must be worn when they are used, but all spray chemicals should be treated with caution.

(C) Fungicides

The fungi are heterotrophic plants which obtain their carbohydrates and protein by digesting the tissues of other plants or animals, usually after these have died. When they do this they are called saprophytes and their waste clearance is vital to continued life on earth.

Fungi which digest plant material while the plant is still alive are parasitic, and will reduce the yield of a crop or perhaps kill it.

Fungicides have been used for years to control blight on the potato plant and more recently to control leaf mildew on barley.

The advent of 'blueprint' husbandry for certain crops, which can lead to very high yields per hectare, has encouraged farmers to control low levels of infection which would formerly have been tolerated. A careful watch is maintained on the crop throughout the year and diseases which are spotted in their early stages can be chemically controlled before they have time to lower potential yields. Crop production has become more scientific and although still very dependent upon the weather it is less subject to the effects of pests and disease.

(D) Reliance upon chemicals

A disturbing feature of the increasing use of chemicals in farming is the apparent increasing resistance of pests to the effects of particular chemicals. When one disease or weed has been effectively controlled, other diseases or weeds seem to increase their competition with the crop plant. Nature adapts, sometimes very quickly, to a change in the environment. For this reason it is important to remember that farming before chemicals was based on a series of sound husbandry practices which should still be applied whenever possible.

Sensible rotation of crops in a field reduces the build-up of disease and improves soil fertility. Ploughing land effectively buries parasites of livestock and crops and helps to remove them from circulation. Varying the varieties grown on the farm and in the country as a whole, reduces the possibility of a rapid spread of disease since each variety has a different degree of resistance. If we continue to rely on sound husbandry we shall have less need to resort to the use of chemicals.

List of supplementary reading

1   Basic principles

Mechanics
Culpin, C. (1981) *Farm Machinery* (Appendix 1, 2, 3). Granada.

Electricity
Basford, L. and Pick, J. (1966) *Lightning in Harness*. Sampson and Low, Maston and Co.

Heat
Stone, R. and Bronner, N. (1962) *General School Physics* (Chapters 6 and 8, Heat and Light). English University Press.

Housing
Farmbuildings. (1982) Noten, Reading University.
Duckham, A. *gen. ed*. (1963) *Farming, Vol. 3* (Chapter 2). Caxton.

Chemistry
Comber and Willcox, *Introduction to Agricultural Chemistry*.
Hay, R. K. M. (1981) *Chemistry for Agriculture and Ecology*. Blackwell Scientific Publications.

Cytology and inheritance
MacKean, D. G. (1971) *Introduction to Genetics*. John Murray.
Whitehouse, H. L. K. (1973) *The Mechanism of Heredity*. Edward Arnold.

2   The plant

Gill, N. T. and Vear, K. C. (1980) *Agricultural Botany*. Duckworth.
Thomas, J. O. and Davies, L. J. *Common British Grasses and Legumes*. Longmans.

3 The animal

Hammond, J. (1983) *Farm Animals* (Chapter 1). Edward Arnold.
MacKean, D. G. (1982) *Introduction to Biology*. John Murray.
Sainsbury and Sainsbury (1979) *Livestock Health and Housing*. Bailliery Tindall.

4 The environment

Russell, E. W. (1950) *Soil Conditions and Plant Growth*. Longmans.
Salt, B. (1982) *Environmental Science*. Cassell.
Mellanby, K. (1980) *Biology of Pollution*. Edward Arnold.
Grundy, K. (1980) *Tackling Farm Waste*. Farming Press Ltd.

5 Beneficial and harmful organisms

Seaman, A. (1963) *Bacteriology for Dairy Students*. Cleaver Hume Press.
Imms, A. D. (1978) *Outlines of Entomology*. Chapman and Hall.
Stevenson, G. (1970) *Biology of Funghi, Bacteria and Viruses*. Edward Arnold.
Scopes and Ledieu (1983) *Pest and Disease Control Handbook*. British Crop Protection Council.

6 Chemicals in crop husbandry

Cooke, G. W. (1982) *Fertilizing for Maximum Yield*. Granada.
Agricultural Chemicals (Approved Scheme booklet) (1985). Ministry of Agriculture, Fisheries and Food.

Glossary of terms

Acaricide: Tick and mite killer
Acid: A sour chemical containing hydrogen replaceable by a metal
Aerobic: In air
Alkali: A chemical which can neutralise an acid
Alluvial: Deposited by rivers or floods
Anaerobic: Without air
Anatomy: Study of the structure of the body
Anion: Particle carrying a negative electric charge
Anode: Positively charged electrode
Antibiotic: Chemical used to combat bacteria
Asexual: Without any fusion of gametes
Assimilation: The utilisation of food material by plants and animals
Bacteria: Microscopic organisms, many of which cause disease
Base: Metallic chemical
Calorie: Formerly used as unit of heat (1 calorie = amount of heat required to raise the temperature of 1 gram of water from 15°C to 16°C). Now replaced by the joule (J) as a standard. 1 calorie = 4.185 J
Capillary: Thin tube
Carbohydrates: An important food requirement of animals; includes sugars, starch etc.
Castration: Removal or destruction of the male sex glands (testicles)
Catalyst: A substance that speeds up chemical action without being itself changed
Cathode: Negatively charged electrode
Cation: Positively charged particle
Centrifugal force: The force that causes a body revolving round a centre to tend to fly off
Chlorophyll: A green pigment found in plants

Colloid: A non-crystalline semi-solid dispersed in a medium, e.g. gelatine and starch

Combustion: Oxidation of organic tissue

Compound (chemical): A substance composed of two or more elements in definite proportions by weight

Conductor: Material which readily transmits an electric current

Convention: Transport of heat by the upward movement of heated particles in liquids and gases

Cytoplasm: Protein colloid found in cells

Deflocculated: Dispersed into individual particles

Depression (meteorological): Area of low barometric pressure

Desiccant: Drying agent

Drench: Draught or dose administered to animals

Earthing: Connecting electrical equipment to the ground as a safety precaution

Effluent: An escaping liquid

Electrolyte: A substance which when dissolved (usually in water) conducts electricity with an accompanying transfer of matter

Electron: Negatively charged particle found orbiting the nucleus of an atom

Emasculation: Another term for castration (see above), or removing the stamens from flowers

Emulsion: Stable suspension of one immiscible liquid in another

Endothermic: Chemical action absorbing heat

Environment: The surroundings

Enzyme: Organic catalyst

Eutrophication: Chemical enrichment of water leading to excessive plant growth and subsequent decay

Excretion: Expulsion of waste products from the body of an animal

Exothermic: Chemical action producing heat

Fermentation: Slow decomposition brought about by microorganisms or substances of plant or animal origin (enzymes)

Fertilisation: Joining together of gametes

Flaccid: Deflated

Flocculated: Small particles aggregated into larger assemblages

Flotsam: Floating debris

Front (meteorological): Line of separation between masses of air at different temperatures

Fumigation: Disinfecting or purifying with fumes

Fungicide: Fungus killer

Galvanising: Protecting iron or steel with a coating of zinc

Gamete: Sex cell

Gene: Factor affecting inheritance

Germination: Commencement of growth in a dormant seed

Gravity: The attraction exercised by a mass (e.g. the Earth)

Haemoglobin: Substance found in blood cells that transports oxygen

Hardy (plant): Plant that grows in exposed situations

Herbaceous (plant): Non-woody green plant

Herbicide: A weed killer

Heterotrophic: An organism that requires its food carbohydrate and protein in a ready formed state

Heterozygous: An organism that has received both the dominant and the recessive elements of an inheritance factor from its parents

Homozygous: An organism that has received the same inheritance factor from its parents. It is pure breeding for that character

Hormone: Substance produced by glands that stimulates other organs of the body

Humidity: Moisture in the air

Hybrid: Organisms whose parents belonged to different races

Hygroscopic: Readily absorbing water

Insecticide: Insect killer

Insulator: Appliance to prevent leakage of electric current from a conductor, or substance to prevent loss of heat

Ion: An electrically charged particle

Joule: Unit of energy including work and quantity of heat. A joule is the work done when the point of application of a force of 1 newton is displaced through the distance of 1 metre in the direction of the force. Or, a joule would be the quantity of heat produced by 1 watt of power per second

Lactation: Production of milk by a female mammal

Leach: The washing of chemicals out of the earth

Leguminous plant: A plant belonging to the family Leguminosae

Lesion: A wound, or damage to living tissue caused by disease

Mammal: Hair-bearing animal with 'warm blood' and (in the female) glands for providing the young with milk

Meiosis: Cell divisions that halve the chromosome number

Membrane: Thin layer or film of living tissue

Metabolism: The chemical processes taking place in living matter

Metamorphosis: Change of shape

Meteorology: Study of the weather

Miscible: Able to be mixed

Mitosis: Cell division which does not alter the chromosome number

Molecule: Smallest particle of a substance capable of independent existence while still retaining its chemical properties

Morphology: Study of the structure and form of organisms

Motility: Ability to move about

Multinucleate: With many nuclei

Neutron: A particle carrying no electrical charge that forms part of the nucleus of an atom

Newton: The unit force which, when applied to a mass of 1 kilogramme, gives it an acceleration of 1 metre per second per second

Nutrient: Conveying, serving as, or providing food

Osmosis: Diffusion of a solvent through a semipermeable membrane into a more concentrated solution in order to equalise the concentrations on either side of the membrane

Oxidation: Attaching oxygen to a chemical, or removing hydrogen from a chemical

Parasite: An animal or plant living at the expense of a host animal or plant

Pasteurise: Heat treating a liquid (e.g. milk) by raising the temperature and maintaining it for a specified period to reduce the numbers of bacteria

Percolate: To filter, e.g. of water through soil

pH: Acid—alkaline scale

Phloem: Sugar conducting tissue in green plants

Photosynthesis: Manufacture of carbohydrates in green leaves

Pigment: Colouring matter

Propagation: Reproduction

Proteins: Complex substances that are essential constituents of the living cell and form a vital portion of the food intake

Proton: A particle carrying a positive electrical charge that forms part of the nucleus of an atom

Pupate: The passing of an insect into the pupa stage of its growth history

Reduction: Removal of oxygen, or the attachment of hydrogen to another chemical

Respiration: Energy production in living cells

Ruminant: Hoofed animal that regurgitates its food for further chewing

Saprophyte: Plant living on dead and decaying organic matter

Secretion: Substance manufactured and discharged by a gland

Solenoid: Cylindrical coil of wire that behaves like a bar magnet when an electric current is passed through it

Solubility: The ability of a substance to dissolve in a liquid

Solvent: A fluid which dissolves another chemical

Spermatozoon: Male sex cell or gamete

Sterilise: Remove the power to reproduce, or to kill all bacteria

Stoloniferous (plants): Producing the essentially horizontal reproductive stems known as stolons

Subcutaneous: Beneath the skin

Sward: Short, cultivated grass

Symbiont: One organism living in harmony with another organism to their mutual benefit

Tiller: Branch produced from the base of the stem in wheat, grasses etc.

Translocation: Water, mineral and sugar movement in a plant

Translucent: Allows light to pass through

Transpiration: Evaporation from leaves

Turgid: Swollen or inflated

Vector: A disease spreader

Viruses: Minute micro-organisms responsible for many plant and animal diseases

Viviparity: Birth of live young, as opposed to egg-laying

Weathering: Breaking-down of rocks by the action of the weather

Xylem: Water conducting tissue in plants

Zygote: Formed by the union of a male and female sex cell

Index

abomasum, 150
acaricides, common, 304
acidic rock, 205
acidity, soil, 78
age, animal, 180
air
    flow, 67
    pressure, 7, 22, 184
    temperature, 195
albumen, 170
alcohol, sterilant, 246
alumino-silicate, 208
amino-acids, 146
ammonium sulphate, 78
amps, 26
anaemia, 158
aneroid barometer, 187
anemometer, 188
animal breeding, 97
    cells, 88
    classification, 101
antibiotics, 239
antibodies, 238
ants, 276
aphids, 240, 269
aqueous solutions, 74
arteries, 161
artificial insemination, 97, 169
atomic structure, 25, 70
axle, 15
azotobacter, 250

bacteria
    milk, 249
    soil, 216, 249
barometric pressure, 186
basalt, 205
battery, electrical, 42
Beaumont, 183, 231
bees, 276
beetles, 272
biological oxygen demand, 85
bionomial system, 102

blackfly *see* aphids
blackhead of turkeys, 255
blight forecasting, 183
blood cells, 158
    clotting, 165
    vessels, 161
bones, 142
boron, 122, 287
bovine tuberculosis, 238
brucellosis, 238
buildings, hygienic, 243
Bunt, 234
butter, 249
butterflies, 271
butyric acid, 242

calcium, 122, 287
carbohydrates, 115, 146
carbolic acid, 246
carbon dioxide, 79, 116
carotene, 180
cells
    animal, 88, 140
    egg, 170
centipedes, 267
cereals, fungus diseases, 233
cervix, 171
chafers, 276
cheese, 248
chemical solutions, 74
    sterilization, milk, 246
chlorophyll, 116
chromosomes, 168
circuits, electrical, 26, 33, 42
clay, 77, 209, 213, 215
click beetle, 275
clones, 177
clostridia, 242
clouds, 199
club root, 232
coccidiosis, 254
coliform bacteria, 247
colloidal solutions, 76

colostrum, 238
competition, plants, 136
compost, 248
conductors, electrical, 32
convection currents, 67
copper, 287
corpus luteum, 170
corrosion, metals, 83
crane fly, 278
creosote, 247
crop yield, 135

density, 4
disease immunity, 238
ditches, 221
drainage, soil, 218
dry cell, 42
dye reduction test, 244
dynamo, 36

earthworms, 210, 218, 264
eelworms, 264
effluent, 85
egg cells, 170
electricity
    cells, 41
    circuits, 26, 33, 42
    coil, 35
    conductors, 32
    current, 26
    insulation, 32
    magnets, 34
    motors, 38
    power, 27
    supply, 28
    transformer, 39
electrolysis, 40
electrons, 25
embryo, 173
emulsions, 75
energy, measurement, 24
enzymes, 155
epididymis, 169
eutrophication, 85
evaporation, 193
eyespot, 236

fallopian tube, 170
fans, 68
farmyard manure, 299
fat, deposition, 179, 180
fats, 117
feedingstuffs, 146
fermentation, silage, 242
fertilization, plants, 129

fertilizers
    compound, 297
    liquid, 300
    residual value, 301
fibrinogen, 166
flea beetle, 272
fleas, 279
floors, heat loss, 64
follicle, Graffian, 170
food constituents, 146
foot-and-mouth disease, 240
friction, 8
frit fly, 279
fungi, soil, 249
fungus diseases, 231
fuses, electrical, 33

gadfly, 277
genetics, 92
gid (sheep), 258
gley soil, 213
gout fly, 279
Graffian follicle, 170
grain chilling, 56
    drying, 56
    weevils, 242, 273
granite, 205
grasshoppers, 268
gravity, 6
greenbottle fly, 277
greenfly see aphids
green manure, 299
Gulf Stream, 197

haemoglobin, 159
heart, 164
heat
    animal body, 63
    capacity, 47
    insulation, 63, 65
    loss, 62
    measurement, 67
    movement, 48
    pump, 59
herbicides, 137, 302
hermaphrodites, 264
hormone production, 170, 174
horsepower, 24
houseflies, 277
housing, livestock, 59–69
humidity
    housing, 66
    measurement, 190
humus, 77, 216
husk, 261
hybrid vigour, 98

hydraulic press, 23
hydrometer, 5
hygrometer, 190

igneous rock, 205
immunity, disease, 238
inclined plane, 20
inheritance, 97
insecticides, 304
iodine, 246
ionization, 40, 72
iron, 122, 288
irrigation, 183, 194, 222

joints, 142
joule, 10, 47

kidneys, 164

leaf area, 135
    blotch, 235
leatherjacket, 278
legumes, 251
levers, 13
limestone, 206
liming, 290
liquid fertilizers, 300
liquids, miscible, 75
liver, 164
liver flukes, 256
locusts, 269
lung, 164

magnesium, 122
magnetism, 33
manganese, 122, 287
mastitis, 237
meat quality, 179
meiosis, 168
Mendel, 92
metamorphic rock, 208
methiocarb, 283
mildew, crops, 191
milk
    preservation, 243
    hygiene, 245
millepedes, 267
minerals, absorption, 118
miscible liquids, 75
mites, 281
mitosis, 90
mole draining, 222
molybdenum, 122
moments, principle see torque
motors, electric, 38
moths, 271
moulds, 230

muscle development, 179

nitrate, 250
nitrogen, 116, 122, 291

Ohm's Law, 26
omasum, 150
oolitic limestone, 206
osmosis, 119
ovary, 170
oviduct, 170
oxidation, 81
oxygen, 80

parallelogram of forces, 8
parasites, 236
pasteurisation, milk, 244
pathogens, 236
peat, 220
penicillin, 239
pH, 82, 215, 290
phenol, 246
phenothiazine, 262
phosphate, 116, 122, 288, 293
photosynthesis, 109, 115, 135
placenta, 173
plant
    classification, 101
    nutrients, 286
plasma, 158
plate count, milk, 244
podsol, 213
pollination, 127
potassium, 122, 287, 295
potato blight, 231
    root eelworm, 263
pressure
    air, 7, 22, 184
    liquid, 23
proteins, 87, 117, 146, 154
pulleys, 12

quartz, 205

rainfall, 191
rechargeable battery, 45
reduction division, 168
red-water fever, 255
refrigeration, 57
reticulum, 150
resazurin test, 243
rinses, milk, 245
roofs, insulation, 65
rumen, 150

saliva, 148, 152

sand, 213
saprophytes, 229
sawfly, 276
sedimentary rock, 208
seed
    dispersal, 130
    dressing, 234
    leaves, 131
seed-borne diseases, 234
septic tanks, 247
semen, farm animals, 170
sewage, 246
sex
    cells, 167
    linkage, 95
sheep strike, 277
silage, 242
silica, 205, 208
silt, 213
skin, 143
slate, 208
slugs and snails, 282
slurry, 300
smut of wheat, 234
sodium, 288
sodium hypochlorite, 246
soil
    acidity, 78
    analysis, 289
    classification, 214
    nutrients, 288
    pH, 82
    profile, 211
    temperature, 196
solenoid switch, 36
spermatozoa, 167
    survival, 171
staggers, sheep, 258
storage, plant products, 241
sugar beet yellows, 241, 288
sugar translocation, 117
sulphonamide, 254
sulphur, 122
sunshine, 115, 198
swabs, milk, 245
systemic insecticides, 271

take-all, 102, 235
tapeworms, 258
temperature
    air, 195
    soil, 196
testes, 167
tetanus, 237
thermometers, 51
thiabendazole, 262

thrips, 268
ticks, 280
toxins, 237
trace elements, 122
transformer, 39
translocation
    sugar, 117
    water, 119
transpiration, 119
trypanosomiasis, 255, 277
tsetse fly, 277
turnip fly, 272
twinning, 173

undulant fever, 238
urethra, 169
uterus, 171

vascular tissue, 107
vas deferens, 169
veins, 163
ventilation, housing, 67
viruses, 240, 270
virus pneumonia, 240
volts, 26
volume, 3

walls, insulation, 65
warble fly, 277
wasps, 276
water
    absorption, 118
    movement, 119
    pollution, 85
    purification, 83
    softening, 84
    supply, 83, 224
    table, 218
    translocation, 119
watts, 27
weather recording, 200
weedkillers, 137, 302
weeds, 136
weevils, 273
weight, 2, 7
wheat bulb fly, 278
wheel, 15
wilting of plants, 220
windbreaks, 189
wind measurement, 188
wireworm, 274
work, 10
worms, parasitic, 256

zinc, 122, 287
zoospores, 231